EUROPEAN INTEGRATION AND FOREIGN DIRECT INVESTMENT IN THE EU

Korean consumer electronics firms have, in the past decade, displayed an exceptional attitude towards direct investment in the EU, marking a definite break from the long tradition of strong governmental initiatives previously undertaken in the strategic affairs of Korean firms. This study addresses the fundamental questions of why and under which conditions Korean firms increased their investment so heavily in the EU after 1986 and why they and not the Korean government took the initiative in the decision-making process.

Dr Sang-Hyup Shin contends that the main reason for the firms' departure from government policy is the belief of the Korean government that national competitiveness, which largely affects the country's standard of living, is determined by the success of local firms in global competition. The government policy of offering tax breaks to Korean *chaebols* is one of the most attractive incentives to expand production abroad rather than at home. Moreover, the government acknowledges that Korean firms are obliged to sell their products in foreign countries because of the small size of the domestic market, which gives rise to the paradoxical situation of barriers being raised to exports, which then, in turn, reduce the competitiveness of firms in the international market.

Dr Shin substantiates his argument by supplying comprehensive surveys of 26 Korean consumer electronics firms and concomitant interviews with senior managers with nine of these. The empirical material suggests that Korean consumer electronics firms were principally motivated by their need to seek market security through a 'globalisation' strategy. Thorough and original, this study addresses key issues in international political economy and will be of interest to corporate strategists and students of international business alike.

Dr Sang-Hyup Shin is senior lecturer at the Department of International Trade and Co-operation, The Graduate School of Pan-Pacific International Studies at Kyung Hee University, Korea.

ROUTLEDGE STUDIES IN INTERNATIONAL BUSINESS AND THE WORLD ECONOMY

1 STATES AND FIRMS
Multinational enterprises in institutional competition
Razeen Sally

2 MULTINATIONAL RESTRUCTURING, INTERNATIONALIZATION AND SMALL ECONOMIES
The Swedish case
Thomas Andersson, Torbjörn Fredriksson and Roger Svensson

3 FOREIGN DIRECT INVESTMENT AND GOVERNMENTS
Catalysts for economic restructuring
Edited by John H. Dunning and Rajneesh Narula

4 MULTINATIONAL INVESTMENT AND ECONOMIC STRUCTURE
Globalization and competitiveness
Rajneesh Narula

5 ENTREPRENEURSHIP IN A GLOBAL CONTEXT
Edited by Sue Birley and Ian MacMillan

6 THE GLOBAL STRUCTURE OF FINANCIAL MARKETS
An overview
Edited by Dilip K. Ghosh and Edgar Ortiz

7 ALLIANCE CAPITALISM AND GLOBAL BUSINESS
John H. Dunning

8 MULTINATIONAL ENTERPRISES FROM THE NETHERLANDS
Edited by Roger van Hoesel and Rajneesh Narula

9 COMPETITION, GROWTH STRATEGIES AND THE GLOBALIZATION OF SERVICES
Real estate advisory services in Japan, Europe and the United States
Terrence LaPier

10 EUROPEAN INTEGRATION AND FOREIGN DIRECT INVESTMENT IN THE EU
The case of the Korean consumer electronics industry
Sang-Hyup Shin

EUROPEAN INTEGRATION AND FOREIGN DIRECT INVESTMENT IN THE EU

The case of the Korean consumer electronics industry

Sang-Hyup Shin

ROUTLEDGE

London and New York

First published 1998
by Routledge
11 New Fetter Lane, London EC4P 4EE

Simultaneously published in the USA and Canada
by Routledge
29 West 35th Street, New York, NY 10001

Typeset in Garamond by Pure Tech India Ltd, Pondicherry
Printed and bound in Great Britain by Biddles Ltd, Guildford and King's Lynn

British Library Cataloguing in Publication Data
A catalogue record for this book is available from the British Library

Library of Congress Cataloging in Publication Data
Shin, Sang-Hyup, 1960–
European integration and foreign direct investment in the EU: the case of the Korean consumer electronics industry / Sang-Hyup Shin,
p. cm. — (Routledge studies in international business and the world economy; 10)
Includes bibliographical references and index.
1. Electronic industries–Korea (South) 2. Electronic industries–European Union countries. 3. Corporations, Korean–European Union countries. 4. Korea (South)–Foreign economic relations–European Union countries. 5. European Union countries–Foreign economic relations–Korea (South) I. Title. II. Series.
HD9696.A3K778 1998
336.4′7621381′095195–dc21
98–20420 CIP

ISBN 0–415–16303–X

TO MY MOTHER AND MY LATE FATHER WITH LOVE

CONTENTS

FIGURES

TABLES

TABLES

PREFACE

This book examines the exceptional nature of the investment of Korean consumer electronics firms in the EU. The main question focuses on the reasons for and the conditions under which Korean consumer electronics firms increased their investment so heavily in the EU after 1986 and why they and not the Korean government took an initiative in the investment decision-making process. This is a break from the long tradition of strong governmental initiatives previously undertaken in the strategic affairs of Korean firms.

This study contends that the main reason for this is that the Korean government acknowledges that national competitiveness, which largely affects the standard of living of the people, is determined by the success of locally based firms in global competition. In addition, the government acknowledges that Korean firms are obliged to sell their products in foreign countries because of the small size of their domestic market, and excessive involvement in markets tends to result in barriers being raised to exports, reducing the competitiveness of firms in the international market. Several leading academic authorities on the subject assert that the national competitiveness of a country is determined by the productivity of firms within that country. However, these explanations are insufficient to explain the Korean case, where the competitiveness of firms in the international market is more relevant and significant due to its heavy dependence on external markets. To maintain or attain the competitive advantage in the world market, Korean firms decided to increase their global economic activities.

To answer the question of why there was such an increase in FDI by Korean consumer electronics firms in the EU, I conducted a survey of 26 Korean consumer electronics firms and interviewed senior managers with nine of these firms. The empirical material thus gathered suggests that Korean consumer electronics firms were motivated by their need to seek market security through a 'globalisation' strategy. The interviews showed that one of the factors determining choice of location for local production within the EU was the various investment incentives offered by host governments in the EU.

PREFACE

There is a general lack of knowledge about firms from less-developed countries (LDCs) – such as Korean firms – and this limits adequate understanding of changes in the world market economy. In order to understand the web of world economy one must be aware of the strategies of the companies involved in global economic activities and why they invest in foreign countries. Therefore, research on FDI by Korean consumer electronics firms in the EU is of importance as it can contribute to more profound understanding of world economic activities where economic interdependence takes place both within the developed world and between the developed and developing worlds.

ACKNOWLEDGEMENTS

I owe a heavy debt of appreciation to all the individuals who have helped in various ways in the preparation of this book. In particular, I should like to express my thanks to Professor Susan Strange at the University of Warwick and Professor Paul Taylor and Dr Michael Hodges at the London School of Economics and Political Science for encouraging me to finish this book.

I wish to express my greatest gratitude to Dr Choue Young-Seek, Chancellor of Kyung Hee University in Korea, and Mrs Choue, both of whom always encouraged me in my research. My personal debt to them is hard to describe. I am also indebted to Professor Han Pyo-Wook, former Ambassador to the UN and the UK, Professor Lee Han-Been, former Deputy Prime Minister and Minister of the Economic Planning Board in Korea, Professor Sohn Jae-Shik, former Minister of the National Unification Board in Korea and Professor Kim Soo-Kon at the Kyung Hee University. Their teaching and support enabled me to finish this book.

I would like to extend my thanks to a number of friends and colleagues who assisted me without hesitation and to all the interviewees who gave so freely of their time to respond at length to my questions.

Finally, I would like to express my deepest gratitude to my mother, brothers and sisters for their endless love, encouragement and support.

Any shortcomings in the research and writing of this book are, of course, mine alone.

Sang-Hyup Shin
Kyung Hee University
Seoul

ABBREVIATIONS

This list is not exhaustive, but covers those abbreviations most frequently used and those which may be unfamiliar to some readers.

ACP	Africa, the Caribbean, the Pacific
ASEAN	Association of Southeast Asian Nations
BOK	Bank of Korea
CEPR	Centre for Economic Policy Research
COREPER	Committee of Permanent Representatives (EU)
CTV	Colour television
CVD	Countervailing duty
DJP	Democratic Justice Party
EAGGF	European Agricultural Guidance and Guarantee Fund
EDCF	Economic Development Co-operation Fund (Korea)
EDP	Economic Development Plan (Korea)
EEA	European Economic Area
EFTA	European Free Trade Association
EIAK	Electronic Industries Association of Korea
EPB	Economic Planning Board (Korea)
ERDF	European Regional Development Fund
ESF	European Social Fund
ESPRIT	European Strategic Programme for R&D in Information Technology
EUREKA	European Programme for high-technology R&D
FECA	Foreign Exchange Control Act (Korea)
FECR	Foreign Exchange Control Regulations (Korea)
FOF	Foreign-owned firm
GATT	General Agreement on Trade and Tariffs
GLIO	Government-led Internal Organisation
GNP	Gross National Product
GSP	Generalised System of Preferences
HCI	Heavy and Chemical Industries
IBJ	Industrial Bank of Japan

ILO	International Labour Organisation
JETRO	Japan External Trade Organisation
KEIC	Korea Export Insurance Corporation
KFTA	Korea Foreign Trade Association
KOTRA	Korea Trade Promotion Corporation
LDC	Less-developed Country
LLDC	Non-ACP Less-developed Country
LIBOR	London Inter-bank Offered Rate
M&A	Merger and Acquisition
MFA	Multi-fibre Arrangement
MFN	Most-favoured Nation
MNC	Multinational Corporation
MNE	Multinational Enterprise
MOF	Ministry of Finance (Korea)
NDRP	New Democratic Republican Party
NIC	Newly Industrialised Country
NIF	National Investment Fund (Korea)
NTB	Non-tariff Barrier
NTM	Non-tariff Measure
OECD	Organisation for Economic Co-operation and Development
PLIO	Privately led Internal Organisation
POSCO	Pohang Iron and Steel Company
PPD	Party of Peace and Democracy
RACE	Research and Development in Advanced Communications Techniques for Europe
RDP	Reunification Democratic Party
SEM	Single European Market
SITC	Standard International Trade Classification
SOK	Korean measurement for rice, equal to a bushed or 80 kg
TRIM	Trade-related Investment Measure
UNCTAD	United Nations Conference on Trade and Development
UNTCOK	United Nations Temporary Commission on Korea
VER	Voluntary Export Restraint

1

KOREAN ECONOMIC DEVELOPMENT AND THE IMPORTANCE OF THE EU MARKET TO KOREAN CONSUMER ELECTRONICS FIRMS

From 1986 to 1994, the total accumulated amount of investment from Korean firms in the EU[1] increased around twelvefold (Bank of Korea 1995). This rate is about three times higher than the average increase in Korean FDI in the rest of the world. The number of Korean FDI projects in the EU has also increased rapidly, especially since late 1986.[2]

To investigate why Korean firms invested so heavily in the EU, especially after 1986, section 1.1 explains chronologically the economic development strategy of the Republic of Korea since the birth of the First Republic in 1947. Section 1.2 concentrates in more detail on the extent of, and the reasons for, the EU's importance to Korean firms.

1.1 Industrial policy and economic development in Korea

Since the 1960s, the Republic of Korea has achieved remarkable economic growth. As Table 1.1 shows, the Korean economy grew at an average annual rate of about 8.7 per cent from 1962 to 1995. The per capita GNP also increased from US$ 87 in 1962 to US$ 10,076 in 1995 at current prices, even though there was a critical period in 1980.[3]

However, as the result of the export-oriented development strategy, Korea is heavily dependent on external markets.[4] Several internal problems of Korean trade made the Korean economy more vulnerable to such external market factors as world-wide protectionism.

Korean policy-makers adopted this export-oriented development strategy because they became aware that import-substitution industrialisation, which the government had adopted in the late 1940s, would reach exhaustion point within the limited Korean market and that growth would have to be

Table 1.1 Major indicators of Korean economic growth (selected years 1962–95)

	A	*B*	*C*	*D*	*E*	*F*
1962	87	2.31	2.2	4.8	−335	130.0
1963	100	2.72	9.1	9.1	−410	130.0
1964	103	2.88	9.6	8.3	−245	256.0
1965	105	3.01	5.8	5.7	−240	272.1
1966	125	3.67	12.7	9.1	−430	271.5
1969	210	6.63	13.8	12.9	−922	304.5
1970	243	7.99	7.6	12.2	−922	316.7
1971	285	9.37	9.1	11.2	−1,046	373.2
1972	316	10.57	5.3	15.1	−574	398.9
1973	396	13.50	14.0	19.8	−566	397.5
1977	1,028	37.43	10.7	22.3	−477	484.0
1978	1,406	51.96	11.0	24.1	−1,781	484.0
1979	1,662	62.37	7.0	22.4	−4,396	484.0
1980	1,589	60.30	−4.8	18.0	−4,384	659.9
1981	1,719	66.20	6.6	17.9	−3,628	700.5
1984	2,044	82.40	8.4	23.2	−1,036	827.4
1985	2,150	83.70	5.4	23.7	−19	890.2
1986	2,300	95.30	12.3	27.8	4,206	861.4
1987	3,218	133.4	12.8	30.8	7,659	792.3
1988	4,295	179.8	12.2	31.5	11,445	684.1
1989	5,210	220.4	22.6	28.4	4,597	679.6
1990	5,883	251.8	14.2	27.4	−2,004	716.4
1991	6,757	292.0	20.0	28.3	−6,980	760.8
1992	7,007	305.7	4.7	27.1	−2,146	788.4
1993	7,513	330.8	8.2	26.7	1,860	808.1
1994	8,508	378.0	14.3	26.0	−3,145	788.7
1995	10,076	451.7	19.5	26.0	−4,747	774.7

Sources: *The Rise of the Korean Economy* (Song 1990: 60–1); *Major Statistics of Korean Economy* (National Statistical Office, Republic of Korea 1996).

Notes: *A*: GNP per capita (US$), *B*: GNP (billion US$). *C*: GNP growth rate (%). *D*: Rate of private savings (%). *E*: Trade balance (million US$). *F*: Foreign exchange rate (Korean won/US$). GNP and GNP per capita are at current prices

based on an external thrust in the early 1960s. Because of heavy reliance on the external markets, world-wide protectionism caused serious problems for Korean firms. In fact, Korean firms have been facing increasingly adverse market situations in developed countries, particularly since the mid-1970s (Nam 1993: 197). Of Korean exports going to 19 developed countries, as Table 1.2 shows, the share of exports that were subject to import restrictions increased from 27.8 per cent in 1976 to a peak of 45.8 per cent in 1981.

Table 1.2 Korean exports under non-tariff barriers (NTBs) by major trading partners (US$ million)

	1976	*1981*	*1985*	*1987*	*1989*	*1993*	*1995*
Exports to the US							
Total exports (A)	2,493	5,661	10,754	18,311	20,639	18,138	24,131
Exports under NTBs (B)	935	2,412	4,656	4,855	4,072	4,506	6,859
Share (= B/A) (%)	37.5	42.6	43.3	26.5	19.7	24.8	28.4
Share (%) of exports under NTBs to other industrial countries:							
Canada	39.3	45.6	31.1	42.9	23.9	20.3	14.3
EU	31.6	39.3	29.5	41.0	22.3	24.9	31.8
Japan	14.5	48.5	32.0	19.4	23.7	5.2	2.7
19 developed countries (C)	27.8	45.8	36.6	31.0	22.3 (D)	18.6 (D)	21.3 (D)

Source: *Overview on Import Restrictions of Major Industrialised Countries* (KFTA 1976: 81, 85, 87, 89, 97).

Notes: NTBs here include voluntary export restraints (VERs), anti-dumping (AD), countervailing duties (CVD), safeguard actions, and other import restrictions under administrative or unfair trade regulations in force or under investigation C: The 19 developed countries are the United States, Canada, Japan, Australia, New Zealand, Germany, the United Kingdom, France, the Netherlands, Belgium, Luxembourg, Denmark, Italy, Greece, Ireland, Finland, Norway, Sweden and Austria. D: Of the 19 developed countries, New Zealand is omitted.

However, the share has decreased since then, to 18.6 per cent by 1993. The declining trend lasted to the end of 1993, although the share increased slightly in 1995. As of the end of 1995, 20 countries in the world set up various types of restrictions on imports from Korea.[5] The export volume that was made under the various restrictions in 1995 amounted to US$ 12,776 million. This was 10.2 per cent of the total Korean export of US$ 125 billion in 1995.

Korean firms suffered from world-wide protectionism because of the nature of their trade. First, Korea's export markets have been excessively concentrated in a few industrially developed countries, especially the USA and Japan, whose extra protectionist measures, as shown in Table 1.2 and Table 1.3, were especially damaging.[6]

As a result of protectionist measures towards Korean exports, the share of Korean exports to the USA under restrictions increased from 37.5 per cent in 1976 to a peak of 43.3 per cent in 1985 (see Table 1.2), although the share decreased to 19.7 per cent in 1989. As Nam (1993: 198) explained, the recent decline in the export coverage of NTBs may be attributed more to the rapid growth of Korean exports than to the increased liberalisation of import restrictions of the United States. In fact, as shown in Table 1.2, Korea's overall exports to the United States increased more than fourfold, from US$ 5.7

Table 1.3 Main trading partners of Korea (selected years 1980–95)

	Share in merchandise exports (%)				*Share in merchandise imports (%)*			
	1980	*1985*	*1990*	*1995*	*1980*	*1985*	*1990*	*1995 (A)*
USA	26.4	35.6	30.3	19.3	21.9	20.8	24.3	22.5
Japan	17.3	15.0	19.4	13.6	26.2	24.2	26.7	24.1
EU (12)	15.5	10.7	13.7	13.0	7.3	9.8	12.1	13.4
of which Germany (B)	5.0	3.2	4.4	4.8	2.9	3.1	4.7	4.9
UK	3.3	3.0	2.7	2.3			1.8	1.8
France	1.7	1.0	1.7	1.8			1.8	1.4
Netherlands	12.0	1.1	1.5	1.2			6.9	8.2

Sources: *Trade Policy Reviews, Korea* (GATT 1992a, Vol. 1, p. 34), *Economic Statistics Yearbook* (Bank of Korea 1997): *Ju Yo Mu Yurk Dong Hyang Ji Pyo* [Major Trading Index] (Korea International Trade Association 1997).

Notes: A: The figures in 1995 mean the percentages of 19 EU member countries. B: Excluding the territory of the former German Democratic Republic. (However, the figure in 1995 includes the territory of the former German Democratic Republic.)

billion to US$ 24.1 billion for the period from 1981 to 1995, while its exports to the United States under import restrictions increased about threefold, from US$ 2.4 billion to US$ 6.8 billion for the same period. Nam (ibid.) gave another reason for the decline of Korean exports to the United States under NTBs. According to him, the decline of Korean exports to the United States under NTBs, particularly during the second half of the 1980s, could have been induced partly by a US policy change. The American government attached more importance to increasing access to foreign markets rather than continuing to increase protection of domestic industries. In recent years, the United States has launched a number of 'Super 301 clause' investigations mainly in order to increase its access to Korean markets. They began with cases for the liberalisation of the insurance market and the protection of US intellectual property rights in Korea in 1985 (ibid.).

Second, as Tables 1.4 and 1.5 show, Korean companies were beset by 20 per cent nominal wage increases, unprecedented labour unrest and appreciation of the won,[7] the Korean currency, over the two years 1988 and 1989. All these internal problems reduced the export-competitiveness of Korean firms in the international market where protectionism was prevalent. Before 1987, South Korea hardly ever experienced strikes. The numbers of strikes in 1987, 1988 and 1989 were 3,749, 1,873 and 1,616, respectively. The average duration of strikes in 1987, 1988 and 1989 were 5.4 days, 13 days and 18 days, respectively (*Financial Times* 16 May 1990). Since 1987, there have been changes in labour–management relations in Korea. Unlike labour movements in other industrial nations, which started and continued as economic entities searching

Table 1.4 Average wages and average wage increases in Korea (1973–92)

Year	*Average*	
	Wages (won)	*Increase rate (%)*
1973	26,954	11.5
1974	35,542	31.9
1975	46,019	29.5
1976	62,362	35.5
1977	82,355	32.1
1978	111,202	35.0
1979	142,665	28.3
1980	176,058	23.4
1981	212,477	20.7
1982	245,981	15.8
1983	273,119	11.0
1984	296,907	8.7
1985	324,283	9.2
1986	350,966	8.2
1987	386,536	10.1
1988	446,370	15.5
1989	540,611	21.1
1990	642,309	18.8
1991	754,673	17.5
1992	869,284	15.2

Source: *Major Statistics of Korean Economy* (National Statistical Office, Republic of Korea 1993: 124).

Table 1.5 Occurrence of labour disputes (1973–94)

Year	*Total*		
	No. of disputes	*No. of workers involved*	*No. of working days lost*
1973	—	—	—
1974	—	—	—
1975	133	10,256	13,557
1976	110	6,570	17,046
1977	96	7,975	8,294
1978	102	10,598	13,230
1979	105	14,258	16,366
1980	407	48,970	61,269
1981	186	34,586	30,948
1982	88	8,967	11,540
1983	98	11,100	8,671
1984	113	16,515	19,984

Table 1.5 (contd)

Year	*Total*		
	No. of disputes	*No. of workers involves*	*No. of working days lost*
1985	265	28,652	64,343
1986	276	46,941	72,025
1987	3,749	1,262,285	6,946,935
1988	1,873	293,455	5,400,837
1989	1,616	409,134	6,351,443
1990	322	133,916	4,487,151
1991	234	175,089	3,257,621
1992	235	105,034	1,527,612
1993	144	108,577	1,308,326
1994	121	104,339	1,484,368

Source: *Major Statistics of Korean Economy* (National Statistical Office, Republic of Korea 1993: 124; 1996: 140).

for economic gains for labour, the labour movement in Korea started as an anti-colonial movement and was maintained as a political tool.

Even during the period of substantial economic growth in the 1960s, 1970s and the first two-thirds of the 1980s, the Korean government used labour as a tool for its economic growth strategy under the policy of 'growth first and distribution later' (Chung and Lie 1989: 230). The Korean government suppressed the labour movement in order to avoid industrial unrest and maintain low production costs. However, the situation changed after the political reform process that started in the summer of 1987. The context of political liberalisation was conductive to work stoppages and strikes in the absence of trade union traditions. Labour aggressively demanded its share of Korea's economic gains and the freedom to engage in collective bargaining. These demands brought tumultuous labour unrest in Korea (ibid.). The major reason for labour disputes was disagreement between labour and management over the rate of wage increases. In 1987, 2,613 cases out of a total of 3,749 cases of labour dispute occurred because of disagreement on wage increases (National Statistical Office, Republic of Korea 1993: 134). As a result, as Table 1.4 shows, the average rate of wage increases is much higher than in the period before 1987.

To show how and why the Korean government adopted an export-oriented development policy, and how and why Korean trade relations have been concentrated on a few industrialised countries, the historical background of the Korean economy is explained here.

The Republic of Korea is comparatively small in area, is resource- and land-poor, and densely populated,[8] like the other three Asian NICs – Hong Kong, Singapore and Taiwan. The economic development of the Republic of Korea following independence from Japan in 1945 may be somewhat arbitrarily classified by political eras. Critical economic policies received different emphases under each regime. Thus, the economic development of the Republic of Korea may be divided into five distinct phases: the typical colonial dual economy under Japanese rule; the period of import substitution and drive for foreign aid under the First and Second Republics (Rhee Syng-Man and Chang Myon); rapid export growth through the export-oriented development strategy under the Third and Fourth Republics (Park Chung-Hee); restructuring, stabilisation and the economic boom (1986–8) of the Korean economy under the Fifth Republic (Chun Doo-Hwan): and a period of adjustment since 1989 under the Sixth Republic (Roh Tae-Woo).

1.1.1 *The typical colonial dual economy under Japanese rule (1910–45)*

The Republic of Korea was originally an agricultural country and until the twentieth century there was practically no modern industry in Korea, as Table 1.6 shows. At that time, necessities of life were produced mostly by household industries of the medieval type. Mining and lumber concessions were granted to America, England and Russia in 1895. In 1899, the first railway in Korea was built by the United States between Inchon, the second-largest port, and Seoul, the capital.

Table 1.6 Percentage distribution of employment in Korea and Japan

	Japan (1878–80)	*Korea (1910)*
Agriculture	83	84
Manufacturing	5	1
Services	12	15

Source: *Growth and Structural Change in the Korean Economy, 1910–1940* (Suh 1978: 37).

However, Western nations had little opportunity to develop industries in Korea because of the Japanese annexation of the country in 1910. Under the protectorate, Japan started to control Korean foreign relations in 1905. Abandoned by the United States under the Taft–Katsura Agreement, Korea was taken by Japan as a result of the Russo-Japanese War of 1904–5, and a Japanese protectorate was established in 1905.

During the colonial period, as Table 1.7 shows, the Korean economy was considerably developed by Japan. However, this development was considerably uneven in character because all decisions on Korean economic development were made to suit Japanese economic needs (Lewis 1988: 6;

Table 1.7 Percentage distribution of net commodity product by industrial origin (1910–40) (based on 1978 values)

Annual average	*Agriculture & forestry*	*Fishery*	*Mining*	*Manufacturing*	*Total*
1910–15	92.5	2.3	1.6	3.6	100.0
1916–21	87.9	3.1	1.7	7.3	100.0
1922–27	85.0	4.1	1.3	9.6	100.0
1928–33	79.5	5.2	2.3	13.0	100.0
1934–39	67.8	5.8	6.4	20.0	100.0
1940	59.6	7.7	8.8	23.9	100.0

Source: *Growth and Structural Changes in the Korean Economy, 1910–1940* (Suh 1987: 16, Table 17).

Table 1.8 The production, consumption and export of rice (1912–36)

	Average production		*Average consumption in Korea*		*Average exported*	
	Million SOK	*Index*	*Million SOK*	*Per Capita SOK*	*Million SOK*	*Index*
1912–16	12.3	100	11.24	0.72	1.06	100
1922–26	14.5	118	10.16	0.59	4.34	409
1932–36	17.0	138	8.24	0.40	8.76	826

Source: *A New History of Korea* (Lee 1984: 348).

Note: One SOK = 180.39 litres (= 5.01 bushels).

Macdonald 1990: 192; Steinberg 1988: 45). In other words, Japanese industrial policy, if there was one, was to develop Korea's industry not as an integral whole but as a complementary structure for Japan's own economy.

As the colonial power from 1910 to 1920, Japan gave priority to the Korean peninsula as a supplier of rice. The export of rice to Japan continued to increase in the 1920s and 1930s, as Table 1.8 shows. In the 1920s, Japanese light industry began to build manufacturing facilities in Korea to take advantage of the nation's low wages and the cheap raw materials in the northern zone of the peninsula. Starting in the 1930s, the Japanese occupation government used the Korean peninsula as an important military base for its expansionism, which facilitated Japan's incursion into Manchuria in 1931 and further into China in 1937.

In order to support its expansionism, Japan began to build a number of large industrial plants, utilising the comparatively abundant resources and raw materials in the Republic of Korea. Most of the heavy industry and power plants were built in the northern provinces where there are abundant mining resources and water power[9] (Provost 1956: 219–24). The southern part of the Korean peninsula saw development only in minor industries such as food processing and the manufacture of cotton textiles. The southern part of the Korean peninsula remained as a kind of resource base to furnish Japan with rice, which accounted for about half of exports. Industries producing non-durable consumer goods were discouraged, however, because Japan wished to conserve market outlets in Korea for its own industries (Balassa 1985: 142; Economic Bureau, Ministry of Foreign Affairs, Republic of Korea 1963: 18). The result of these Japanese policies was a typical colonial dual economy, in which over half a million Japanese residents managed the modern industrial sector and enjoyed its fruits, while most Koreans – except the urban labour force – remained in the traditional agrarian economy, as Tables 1.9 and 1.10 show.

Table 1.9 Occupational distribution of Korean male workers (1930–40)

	1930		*1940*		
	No. of workers (1,000)	*Per cent distribution*	*No. of workers (1,000)*	*Per cent distribution*	*Difference*
Agriculture	5,019	81.1	4,543	71.5	−476
Fishery	93	1.5	113	1.8	20
Mining	30	0.5	161	2.5	131
Manufacturing	247	4.0	374	5.9	127
Others	801	12.9	1,163	18.3	362
Total	6,190	100.0	6,354	100.0	164

Source: Population census data, Korean government.

Table 1.10 Occupational distribution of Japanese male workers in Korea (1930 and 1940)

	1930 No. of workers (1,000)	*1930 Per cent distribution*	*1940 No. of workers*	*1940 Per cent distribution*
Agriculture	13	7	11	5
Fishery	7	4	6	3
Mining	1	1	9	4
Manufacturing	37	21	65	28
Commerce	34	19	39	17
Transportation and communication	19	10	40	17
Civil service and professional	66	36	60	26
Other	4	2	1	0
Total	181	100.0	231	100.0

Source: *Growth and Structural Changes in the Korean Economy, 1910–1940* (Suh 1978: 54, Table 24).

1.1.2 Import substitution and foreign aid maximisation under the First and Second Republics

This period can be also divided into three: from August 1945 to July 1947, from August 1948 to April 1960, and from May 1960 to May 1961.

During the period from August 1945 to July 1947, the US military administration governed the south of Korea, and the economy in South Korea was in a serious situation because of the partition of the Korean peninsula and the lack of an economic development policy. The Republic of Korea gained independence from Japan in 1945. However, the Republic was divided into two parts, North Korea and South Korea, by the military occupations of the USSR and the USA, respectively.[10] In economic terms partition was a disaster because, as explained above, most industries of Korea were in the north while the bulk of the population and agricultural land was in the south, according to the Japanese need (Ministry of Public Information, Republic of Korea 1966: 29).

During the period from August 1945 to July 1947, there were no economic development policies, including industrial policies, in Korea because the US military administration gave priority to the negotiation of reunification with the Soviets. It has also been said that the administration was prejudiced against centralised state economic planning (Macdonald 1990: 194).

During the period from 1948 to 1960, the First Republic was launched and the Korean War broke out. Only in 1948, three years after independence from Japan, were there elections in the south, on 9 May.[11] On 15 July 1948, Rhee Syng-Man was elected as the first President of the Republic of Korea by the first National Assembly of the country. As the first President, Rhee adopted economic policies he felt would help to develop the country. However, the policies were not effective because they were naive, populist, undisciplined and inflationary (ibid.: 195).

In 1950, only about two years after the birth of the First Republic, there was a historical incident that shattered the fledgling economy of South Korea. The Korean War broke out and lasted for three years. During the Korean War, most infrastructure and industrial capability was severely damaged.[12] In the aftermath of a devastating war, the Republic of Korea was one of the poorest countries in the world, its war-damaged economy based largely on agriculture and massive injections of foreign aid (Korea Development Institute 1975: 14).

The period following the Korean War was characterised by import substitution in non-durable consumer goods and their components. It was during this period that President Rhee Syng-Man adopted the economic policies characterised by import substitution and a drive for foreign aid (Steinberg 1988: 126; Macdonald 1990: 196).

Macdonald described the economic policies of the First Republic after the Korean War as follows:

> President Rhee and his supporters had scant understanding of economics. Their strategy was to maximise foreign aid, overvalue their currency, meet the government deficit by printing money and bonds, keep interest rates artificially low, and focus on import substitution for economic growth.
>
> (1990: 197)

The overall import-substitution policy restricted imports and protected and perpetuated many fledgling and uncompetitive Korean industries. The application of this policy led to the expansion of industries producing non-durable consumer goods and their components. The Korean government put the primary emphasis on food and other consumer goods to meet minimal living requirements, though there was some investment in import-substitution industries and textiles. Import substitution accounted for 24 per cent of industrial growth in the second half of the 1950, while exports contributed only 5 per cent (Westphal and Kim 1982: 258). At the same time, exports were discouraged by overvalued exchange rates,[13] because balance-of-payments equilibrium was maintained at this time through large inflows of foreign aid, and rapid domestic inflation[14] was not fully offset by the depreciation of the won.

The government attempted to reduce the adverse effect on exports of the overvaluation of the Korean won by allotting certificates to the holders of foreign currency which allowed them to import certain popular items, essentially establishing a multiple exchange rate system. However, the measures only compensated to a small extent for the overvaluation of the won, and exports were further discouraged by fluctuations in the real exchange rate that created considerable uncertainty over the domestic currency value of export earnings.

The process of import substitution continued behind high protection, with increasing use made of quantitative restrictions. By 1957, however, possibilities for first-stage import substitution were increasingly exhausted, since domestic production accounted for nine-tenths of the consumption of non-durable consumer goods and their components.

After the war, the shortage of goods, which had already spread over the whole economy, was further aggravated by two factors: insufficient supply due to the limited production capacity, and the sharp increase in consumption due to the explosive influx of refugees from the north during the war.[15] The Korean economy was so strained that the government could not alleviate the shortage of materials without foreign aid. From 1945 to 1961, the Korean government received foreign aid of US$ 3 billion (Korea Development Institute 1975: 14). Aid peaked in the three years from 1956 to 1958, when an average US$ 300 million of foreign aid flowed into Korea annually. The foreign aid brought positive and negative effects to the Korean economy at the same time. The positive function of the foreign aid was to stabilise the economic situation and to channel funds into reconstruction of productive facilities (Economic Bureau, Ministry of Foreign Affairs, Republic of Korea 1963: 20). The negative function was that foreign aid did not contribute significantly to economic development because it consisted of mainly consumer goods; on the contrary, the aid resulted in adverse consequences such as an increase in the national propensity to consume, and inefficiency in government fiscal operations (Korea Development Institute 1975: 14).

In summary, from 1948 to April 1960, the Rhee government implemented import-substitution policies and manipulated the foreign-aid programme on which Korea relied, and the country recorded a poor economic growth rate.[16]

The Second Republic (1960–1) was formed by an interim government in the aftermath of the student revolution that overthrew President Rhee Syng-Man in April 1960. A European-style cabinet form of government was adopted with Chang Myon as Prime Minister.

The Second Republic lasted for only nine months. During this period, the government of Prime Minister Chang Myon did its best to restore the Korean economy. To this end, first, the Chang government devalued the Korean won substantially. The government also adopted unified exchange rates and liberalised import controls. To promote exports, the Prime Minister brought in an export-subsidy scheme which provided various advantages to exporters, such as a 50 per cent tax reduction on profits derived from exports and full exemption from indirect taxes on their inputs as well as on export sales. The government prepared a three-year plan and a five-year plan. These policies and plans, however, proved abortive because of the collapse of the regime.[17] In fact, Prime Minister Chang's government was ineffective largely because of the political bickering and instability that followed President Rhee's resignation. Relations between the government and firms during the period were not firmly established because the private sector was so small.

1.1.3 Rapid export growth under the Third (1961–72) and Fourth (1972–9) Republics

Under the Third and Fourth Republics, the government played a very significant role in the economic development of Korea. The period of the Third Republic and the Fourth Republic can be seen as the 'take-off period' and the 'heavy industry drive period', respectively (Lee and Yamazawa 1987: 55–8).

In 1962, one year after Park seized power, he adopted a series of five-year Economic Development Plans (EDPs). In fact, the First Five-year Economic Development Plan, which President Park started in 1962, was the starting-point for Korea's miraculous economic growth. From 1962 to 1979, the first, second, third and most of the fourth economic plans were implemented.

Through the EDPs, the Korean government adopted an export-oriented development strategy between 1961 and 1979, because the limited domestic market could not provide opportunities for further import substitution – an approach that the government had adopted after the Korean War. Policies were geared towards raising export competitiveness and increasing output of labour-intensive manufactured exports. To promote exports, the government of the Third Republic adopted various policies reflecting the changes in internal as well as external factors[18] and constructed basic industries and a social infrastructure through the First and Second Five-year Economic Development Plans,[19] as Table 1.11 illustrates. In fact, these were the precondition for the promotion of exports.

Through the five-year plans, the government took on the role of guide for the state's industrialisation process. From the First Five-year Economic Development Plan in 1962, the government tied its industrial progress to expansion of exports of manufacturers. Industrial policy in Korea during the period can be divided into three distinct phases of evolution: (a) 1962–4, (b) 1965–73 and (c) 1974–9.

In the first period, from 1962 to 1964, the government continued to pursue an import-substitution policy. Investments were undertaken to replace foreign imports in such areas as cement, oil refining and fertilisers. During

Table 1.11 Summary of the First and Second Economic Development Plans

	First EDP	*Second EDP*
PO	1 Correction of socioeconomic vicious circle of previous years. 2 Establishing the foundation of a self-reliant economy.	1 Modernisation of industrial structure. 2 Promotion of self-reliant economy.
Period	1962–6	1967–71
EGR[a]	7.1 (8.5)	7.0 (9.7)
MP	1 Securing an energy supply source. 2 Correction of structural inbalance in economy. 3 Expansion of key industries and social overhead capital. 4 Effective use of idle resources. 5 Improvement of balance of payments position. 6 Technical development	1 Self-sufficiency of grains and development of forestry and fishery. 2 Formation of foundation for industrialisation. 3 Improvement of balance of payments position. 4 Raising employment ratio and restriction of population increase. 5 Raising farmers' income. 6 Improvement of technology and productivity.

Source: Economic Planning Board, Republic of Korea.

Notes: EDP: Economic Development Plan. PO: Principal objectives. EGR: Economic growth rate. MP: Major policies. a: The figure in parentheses is actual economic growth rate, while the figure without parentheses is target economic growth rate.

this period, the country attained an annual growth rate of 6.9 per cent, but this was at the cost of incurring a substantial degree of inflation. However, the growth strategy based on import substitution soon reached its limits because of small domestic markets, large capital requirements and foreign exchange constraints, which became increasingly serious following a decline in US aid in the early 1960s. Thus, the government adopted an outward-looking industrialisation strategy in the mid-1960s. From 1961 to 1965, in fact, the government greatly expanded the system of export incentives, with a view to removing the anti-export bias inherent in existing import restrictions.[20]

The expanded system of export incentives included preferential export credit, which became one of the most important incentives. A comprehensive drawback system was introduced for tariffs and indirect taxes on raw materials for export production. Direct tax reductions were given on income earned from exporters, and wastage allowances were permitted for raw materials imported for export production. These concessions were available to all exporters. Simultaneously, an export mentality was fostered through a series of actions taken by the government. These actions included the monitoring of informal export targets, and high-level government meetings designed to discuss and resolve the difficulties faced by exporters (these included well-publicised monthly meetings such as those of the Export Promotion Council,[21] which was often chaired by the President). During the early period of expansion, the government kept the exchange rate at a generally realistic level, holding the real effective exchange rate roughly constant (Park 1989; Kim 1991).

The second period, from 1965 to 1973, can best be characterised as one of export-led economic development. During the export-led period, the government encouraged the growth of manufactured exports. The government gave the greatest support to three manufacturing sectors: textiles, clothing and electronics. To promote exports, the government provided exporters with tax exemptions, an almost unlimited supply of credit at a preferential interest rate and a continuing supply of capital as needed. It was possible to supply the above because, as Hong (1993: 424) points out, the Korean government was the single major shareholder in all nation-wide commercial banks by the early 1980s. In addition, it was possible because the Korean government could control firms' access to foreign borrowers. Foreign capital, which reached a cumulative total of US$ 16.2 billion at current prices in the period 1962–79, was in the form of public and commercial loans (93.8 per cent) rather than direct investment (6.2 per cent) (EPB 1980b: 145). The contribution of foreign investment to Korean domestic capital formation was substantial. In fact, foreign investment accounted for more than 30 per cent of annual average investment between 1962 and 1979 (EPB 1980a: 14). Besides these various export incentives, the government devalued its currency successively to encourage exports. Concurrently, the government adopted a relatively liberal import policy. As an example, the list of import items subject to automatic licensing approval increased from some 12 per cent to 60 per cent of the tariff code in the period from 1965 to 1967. In fact, other LDCs, like India, Argentina and Brazil, started the automatic licensing approval system in the 1980s.

During the period of the Third Republic (1961–72), the government also adopted the following three industrial policies in order to promote exports. First, the government changed the exchange rate system to a uniform flexible system in which the won was pegged to the US dollar. Second, the government raised the nominal rate of interest to about 30 per cent to induce increased domestic saving. The nominal rate of interest had been below the inflation rate for some time. Third, the government adopted fiscal and monetary stabilisation policies in order to curb inflation.

During the third period, from 1973 to 1979, the emphasis of industrial policy shifted from export promotion towards promotion of the heavy and chemical industries (HCI) to develop the capital goods industry and inputs of intermediate products. The government encouraged investment in these sectors through industrial incentives such as preferential tax and credit incentives offered by a newly established National Investment Fund (NIF). Thus, the year 1973 was a turning-point for industrial policy in Korea.

Tax incentives that the government devised for these 'important industries' included complete exemption from corporate and income tax for three years, followed by taxation at 50 per cent of normal rates for the subsequent two years (Yoo 1984: 170). The government subsidised the interest rate difference and made loans[22] to these industries at a preferential rate as well. These incentives were offered to such 'important industries' as petrochemicals, shipbuilding, machinery, electronics, steel, non-ferrous metals, fertiliser, defence industry, electric power plants, aircraft and mining activities. In addition to indirect financing through NIF, the government also made substantial direct investments, in the form of infrastructure construction or subscription to important industries.

Through the HCI policy, the government was able to upgrade Korea's industrial structure in a very short time and lay the foundation for some elements of the export growth of the 1980s. However, the HCI policy brought serious structural imbalances to the Korean economy.[23] One of the most serious problems that resulted from the excessive investment in industry was inflationary pressure. As a policy preference, loans were extended to HCI projects at interest rates that were negative in real terms, and an excessive demand for funds developed. With the increase in the money supply, prices rose, workers demanded higher wages and export competitiveness suffered. And when the government devalued the currency in order to restore export competitiveness, the economy experienced a new round of inflationary pressure stemming from the high cost of imports in local currency.

The second serious problem had to do with unbalanced growth among firms and industries. Since the favourable incentives were allocated to investment in HCI industries on a priority basis, investment in other sectors of the economy suffered. In other words, the biased incentive structure, particularly cheap preferential credit combined with moral persuasion to meet investment and export targets, led to substantial overcapacity in some HCI sectors. In addition, industrial concentration increased because only a few industrial groups were able to undertake such projects, which assisted the *chaebols* to become firmly established.[24]

The third problem had to do with the weakened competitiveness of South Korean industry. As part of the effort to accelerate the development of HCI, the government protected these industries excessively. Furthermore, the extensive government controls and regulations sapped private initiative and creativity, since, for example, loans were granted to the HCI as a matter of policy preference. These difficulties were aggravated by an international economic slowdown and a disastrous harvest.

During this period, as Table 1.12 shows, relations between the government and firms can be described as 'brotherhood relations'. The government played the role of a big brother, while firms played the role of a younger brother. By these means, the government was in a good position to guide and control the firms.[25]

Owing to the outward-oriented development strategy during the period of the First and Second EDPs, the Korean government achieved the structural change that transformed a backward, primary-product-based economy into a newly industrialising developing country, as Table 1.13 illustrates.

Table 1.12 Industrial policies and the relations between government and firms in Korea (1945–present)

Period	*Policies*	*Relations between the government and firms*
1945–61	Import-substitution and aid-maximisation policy.	No relations due to the lack of industries.
1962–64	Import-substitution policy	Emergence of the relations between the two parties.
1965–73	Export-led economic development policy.	Government played the leading role.
1973–79	Heavy and chemical industry (HCI) policy.	Government played more leading role. Relations like a big brother and a younger brother.
1980–87	Stabilisation and adjustment policy.	Still strong relations but government's role reduced slightly.
1988–present	Adjustment policy.	Still strong relations, but the firms start to get as powerful as the government.

Table 1.13 Changes in industrial structure (selected years 1962–85)

Industry	*1962 (% employed)*	*1966 (% employed)*	*1972 (% employed)*	*1976 (% employed)*	*1982 (% employed)*	*1985 (% employed)*
A	36.6	34.9	26.4	23.5	14.6	13.8
B	16.2	20.5	23.1	28.4	29.8	29.6
C	25.8	34.1	34.9	45.9	51.2	54.4
D	74.2	65.9	65.1	54.1	48.8	45.6
E	47.1	44.7	50.5	48.1	56.6	56.6

Sources: *Major Statistics of the Korean Economy* (EPB 1980a, 1986); *The Economic Development of Japan and Korea* (Lee and Yamazawa 1987: 56).

Notes: A: Agriculture, forestry and fisheries. B: Mining and manufacturing. C: The ratio of output of heavy and chemical industries to total output of mining and manufacturing. D: The ratio of output of light industry to total output of mining and manufacturing. E: Social overhead capital and other.

The structural change made Korea exploit its comparative advantage in international trade and also contributed to increases in productivity. While adopting an outward-oriented development policy, the Korean government had also sought economic stabilisation by undertaking an emergency currency reform in an effort to curb inflation. This economic stabilisation has been the most urgent task to be performed in the process of economic development since the First Five-year Economic Development Plan was launched in 1962. Although the currency reform was not entirely successful in bringing about the expected results, with this reform as a precedent, the goal of stabilisation of the national economy has been steadily pursued ever since. In 1963 the government also initiated the Financial Stabilisation Programme to curb inflation. In 1965, the adjusted interest rate rose to realistic levels.

In the process of the industrialisation of the Korean economy from 1962 to 1978 Korea was influenced by the USA and Japan.[26] Korean political and economic leaders of the time were clearly strongly influenced by Japan in the formation of Korea's economic development strategies (Kwak, Patterson and Olsen, 1983: 40–1; Lee and Yamazawa 1987: Introduction; Steinberg 1988: 131).

Considering the geographical proximity of Korea to Japan,[27] their historical relations,[28] the spectacular success of the Japanese growth process and the similarities between the two countries, such as the lack of domestic natural resources and the fact that both entered their post-war periods with battle-ravaged economies, it should be easy to see why the Korean government emulated the Japanese model of economic growth more than the USA's model. This was true especially after 1965, when diplomatic relations between Korea and Japan were normalised. When one considers the way Korea obtained the foreign technologies necessary for industrialisation during the period from 1962 to 1980, it becomes clear how significant was the

Japanese influence. In the early stage of industrialisation (1962–72), licensing agreements or foreign direct investment were not the most important ways of acquiring foreign technologies for Korea. Korea relied more on technologies imported through capital goods, contact with foreign buyers and experiences with foreign-owned firms. As Korea was becoming industrialised, however, it began to rely heavily on importing technologies through licensing agreements and foreign direct investments (Lee and Yamazawa 1987: 88). During this period, as Tables 1.14, 1.15 and 1.16 show, Japan was the largest supplier of licences and the largest investor in Korean manufacturing industries.

Table 1.14 Direct US and Japanese investment in the Korean manufacturing industries (1973–9)

Industries	*Low-tech*	*High-tech*	*Total*
Japanese direct investments			
Labour-intensive	60 (18 %)	188 (57 %)	248 (75 %)
Capital-intensive	32 (10 %)	49 (15 %)	81 (27 %)
Total	92 (28 %)	237 (72 %)	329 (100 %)
US direct investment			
Labour-intensive	6 (19 %)	13 (41 %)	19 (60 %)
Capital-intensive	1 (3 %)	12 (38 %)	13 (40 %)
Total	7 (22 %)	25 (78 %)	32 (100 %)

Source: *The Economic Development of Japan and Korea* (Lee and Yamazawa 1987: 91).

Notes: Labour-intensive, high-technology industries: printing, publishing and allied industries; machinery; electrical machinery, apparatus; appliances and supplies; transport equipment; and professional and scientific equipment. Capital-intensive, high-technology industries: industrial chemicals; other chemical products; petroleum products; and basic non-ferrous metal products. Labour-intensive, low-technology industries: food; tobacco, textiles, clothing; leather, leather products, leather substitutes and fur; footwear; wood and wood cork products; furniture and fixtures; rubber products; plastic products; fabricated metal products (except machinery and equipment); and other manufactures. Capital-intensive, low-technology industries: beverage; paper and paper products; pottery, china and earthenware; glass and glass products; other non-metallic mineral products; and basic iron and steel products.

Table 1.15 Direct US and Japanese investment in the Korean manufacturing industries (1962–72)

Industries	*Low-tech*	*High-tech.*	*Total*
Japanese direct investments			
Labour-intensive	40 (48 %)	21 (25 %)	61 (73 %)
Capital-intensive	14 (17 %)	9 (11 %)	23 (27 %)
Total	54 (64 %)	30 (36 %)	84 (100 %)

Industries	*Low-tech*	*High-tech.*	*Total*
US direct investment			
Labour-intensive	14 (39 %)	12 (33 %)	26 (72 %)
Capital-intensive	4 (11 %)	6 (17 %)	10 (28 %)
Total	18 (50 %)	18 (50 %)	36 (100 %)

Note: As Table 1.14.

Table 1.16 Import licences approved by the Korean government (1962–80)

	Japan	*USA*	*Subtotal*	*Others (A)*	*Total*
1962–6	11	13	24	9	33
1967–71	203	61	264	21	285
1972–6	280	90	370	64	434
1977–80	520	227	747	227	97
1962–80	1,014 (58.7%)	391 (22.6%)	1,405 (81.4%)	321 (18.6%)	1,726 (100%)

Sources: *Handbook of Korean Economy* (EPB 1980b); *The Economic Development of Japan and Korea* (Lee and Yamazawa (1987: 93)).
Note: A: Includes West Germany, Britain, France and others.

Because of Seoul's economic development strategies, Korean firms have competed with Japanese firms in the international market since the mid-1970s. Major exporting items of the Republic of Korea are very similar to those of Japanese firms.

During the period of the Third and Fourth Republics (1962–79), as Table 1.1 shows, the average GNP growth rate was about 9.2 per cent, and the GNP per capita increased by 19 times from US$ 87 in 1962 to US$ 1,662 in 1979.

1.1.4 The restructuring and stabilisation of the Korean economy under the Fifth Republic (1980–8)

The period of the Fifth Republic was one of structural adjustment and liberalisation. In 1980, the Fifth Five-year Economic Development Plan was the first to incorporate the concepts of an industrial policy. The government focused on the structural imbalances in the economy. In fact, Korea's recent macro-economic history is essentially a chronicle of the fifth five-year plan, which was initiated in 1982 and revised and relaunched in 1984, as Table 1.17 illustrates. In 1980 the Korean economy had some serious problems caused by internal and external factors. The internal factors were high domestic inflation, poor harvest and political instability following President Park's death. The external factors were the world-wide recession following the

Table 1.17 Summary of the Third, Fourth and Fifth Economic Development Plans (EDPs)

	Third EDP	*Fourth EDP*	*Fifth EDP*
PO	1 Coordination of growth, stabilisation and equity. 2 Achievement of self-reliant economy. 3 Maximisation of land cultivation and balanced regional development.	1 Achievement of self-generating growth. 2 Promotion of equity through social development. 3 Technical innovation and efficiency improvements.	1 Stabilisation of prices. 2 Raising of productivity. 3 Promotion of social development. 4 Elimination of distortions in the economy and promotion of rational development.
Period	1972–6	1977–81	1982–6
EGR[a]	8.6 (10.1)	9.2 (5.5)	7.56 (8.729)
MP	1 Self-sufficiency in food staples. 2 Improvement of living environment in rural and fishing villages. 3 Improvement of balance of payments position. 4 Improvement of industrial structure. 5 Development of technology and human resources. 6 Expansion of social overhead capital. 7 Effective use of resources. 8 Improvement of welfare.	1 Securing domestic investment resources. 2 Achievement of stable balance of payments. 3 Strengthening of international competitiveness. 4 Expansion of employment opportunities and manpower development. 5 Expansion of Saemaul campaigns. 6 Improvement of living conditions. 7 Expansion of investment in science and technology. 8 Improvement of economic management and system.	1 Curbing inflation 2 Recovery of competitive power in heavy industry. 3 Consolidation of agricultural policy. 4 Improvement of financial system. 5 Overcoming energy constraints. 6 Adjusting Government function and rationalisation of fiscal management. 7 Shift to competitive system and promotion of open economy policy. 8 Development of education and manpower and promotion of science and technology. 9 Establishing new relationships between labour and management. 10 Expansion of social development.

Source: Economic Planning Board, Republic of Korea.
Note: As Table 1.11.

second oil shock in 1979, growing protectionism in the advanced countries and the rapid emergence of new strong competitors specialising in the low-skilled and labour-intensive goods that had been the major Korean exports. These internal and external factors made Korea feel the need to change its industrial structure.

During this period, the government adopted five major policies to change its industrial structure. First, the government implemented the macro-economic policy which aimed at stabilising inflation.[29]

Second, the government readjusted the structure of the HCI by initiating mergers, cancelling some investment projects in the industries and promoting product specialisation in power-generating equipment, automobile assembly, heavy electric motors, electronic switchboard systems, diesel engines for marine use, and copper smelting. Most firms in these industries had recorded large losses because of the overlapping and excessive investments in the 1970s and the lack of demand caused by the world-wide recession.

Third, the government revoked the large incentives that were given to HCI sectors in the 1970s. The typical incentives had been loans at preferential rates of interest and tax exemptions.

Fourth, the government promoted market competition and eliminated various factors inhibiting a competitive environment. By promoting market competition, the government sought to enhance efficiency. In order to promote further competition, the government also persistently pursued import liberalisation during this period.

Fifth, the government emphasised growth of small- and medium-sized firms. To this end, the government promoted manufacturing of technology-intensive or skilled labour-intensive products such as colour televisions (CTVs), refrigerators, computers, semiconductors and machine tools.

These efforts to develop the Korean economy were successful. During the nine years from 1980 to 1988, the average growth rate of GNP was 7.8 per cent, as Table 1.1 shows, and 1986 in particular was a landmark year for Korea, which made the most of the three lows: the fall in oil prices, wages[30] and interest rates. The country's current account crisis was alleviated, and, from 1986, the Korean economy entered a phase of greater financial autonomy. Unexpectedly, the current account surplus which appeared in 1986 in the context of the three lows, grew in the context of the three highs (oil prices, won and wages). In 1987, Korean GNP amounted to US$ 118.6 billion while the GNP per capita was US$ 3,098.

As of the end of 1987, the Republic of Korea ranked seventh in the world in the size of its trade surplus, eighth in domestic savings rate, twelfth in total trade volumes (US$ 88.3 billion), fifteenth in automotive production and eighteenth in its size of GNP. In 1988 the economy continued to grow at the annual nominal rate of 12.2 per cent. In 1988, Korean GNP amounted to US$ 169.2 billion and GNP per capita reached US$ 4,040. Total exports jumped to US$ 60.6 billion and imports to US$ 51.8 billion. The trade surplus reached a record of US$ 8.8 billion, as Table 1.1 shows.

1.1.5 *The period of adjustment since 1989 under the Sixth Republic (1989–93)*

Through the Sixth Five-year Economic Development Plan, which was implemented under a policy environment similar to the one that characterised the implementation of the fifth plan, the government continued to adjust its economy during the period of the Sixth Republic. Perhaps the most prominent goal of the sixth plan, which was implemented over 1987–91, was the achievement of a reduction in Korea's net external debt. Concurrently, the government also continued to pursue policies of trade and financial liberalisation. On the trade side, almost 95 per cent of imports were allowed entry without quota restrictions (and with declining tariff protection) by 1991. On the financial side, increasing autonomy was given to commercial banks in their credit pricing and allocation decisions. The financial sector was also internationalised by progressively opening up to foreign banks and securities firms.

After the period of the economic boom from 1986 to 1988, the government encountered internal and external pressures. The high growth rates of the boom period (12.9 per cent in 1986, 13 per cent in 1987 and 12.4 per cent in 1988) strained capacity utilisation and tightened labour markets, generating inflationary pressures. Korea's current account surpluses exacerbated trade tensions. Trade partners called for market opening, further demand expansion and appreciation of the real effective exchange rate. In addition, after the Declaration of Democracy in June 1987, which led to the surfacing of labour demands, the number of working days lost to strikes rose sharply and the upward pressure on wages increased further.

The government reacted to these internal and external pressures by changing policy direction in 1987 towards reducing the current account surplus while sustaining growth and improving living standards in a non-inflationary manner. This policy reorientation brought a sharp adjustment in late 1988 and 1989. In 1988, the pattern of growth began to shift towards satisfying domestic demand, as sizeable wage settlements led to a substantial increase in disposable incomes. The 16 per cent nominal appreciation of the won *vis-à-vis* the US dollar, together with rising inflation and unit labour costs, eroded the competitiveness of Korea's exports. These trends continued in 1989 when wage increases significantly exceeded productivity, the real effective exchange rate appreciated by a further 14 per cent and the volume of exports declined for the first time since 1979, which reduced domestic growth.

However, the slowdown in growth was reversed in 1990. Real GDP increased by 9 per cent and domestic demand rose rapidly, largely owing to the enhanced government efforts to improve the social infrastructure, particularly housing. In 1991 the Korean economy continued to show signs of overheating. Thus, the government took a number of micro-economic measures aimed at curbing inflation and addressing the external deficit. The issuance of new construction permits for certain building categories was suspended from May 1991 to June 1992, and in September 1991 a mandatory delay on commercial construction was imposed. These measures were designed to ease domestic shortages and reduce demand for imported construction materials. Market opening was also being pursued, with the first stage of a new three-year programme to reduce the scope of non-automatic licensing for imports implemented in January 1992. Concurrently, the average tariff was reduced from 11.4 per cent to 10.1 per cent. The first phase of a four-stage interest rate liberalisation plan was also implemented, deregulating the rates on most money instruments.

Under the Sixth Republic, as illustrated in Table 1.12, the private sector started to have a stronger voice in relations with the government, even though the government still held the more powerful position. However, it was clear that the private sector was in a relatively strong position in that the chairman of a *chaebol* criticised government policy publicly, which would previously have been unimaginable. This change came from the large size of *chaebol*s and the dramatic change in overall social circumstances which was further facilitated by the Declaration of Democracy in June, 1987.

1.2 The importance of the EU market to Korean firms

Korean firms experienced difficulties because of their export-oriented development strategy and internal problems such as unbalanced trade relationships, dramatic wage increases in 1988 and 1989, labour unrest and appreciation of the won. Considering this, Korean firms have needed to make every effort to maintain and increase their trade volume in the international market, by improving their competitive edge, and by overcoming internal and external difficulties, and, in particular, to by-pass world-wide protectionism. Korean firms can achieve this and reach foreign markets through FDI. This is a strategy complementary to exports for penetrating foreign markets over time (Thomsen and Nicolaides 1992: 69).

Korean direct investment has increased dramatically since the mid-1980s.[31] By region, Korean investment is concentrated in North America, especially in the USA, which is the major market for export products, and in Southeast Asia, which has abundant manpower and natural resources.[32] Recently, however, Korean investment abroad has begun to diversify into other areas such as Europe, South and Central America and Oceania.[33]

Among these, as an area for investment, the EU is especially important to Korea for the following reasons. First, there is the fear of 'Fortress Europe', that the EU will become inward-looking and more protectionist than ever before. The birth of the Single European Market (SEM) is the final stage in the process of economic integration of trade in goods, services and the free movement of individuals.[34]

European integration can be interpreted differently depending on the situation of the country concerned. The EU countries insist that integration is not a prelude to protectionism. The European Commission, in fact, has taken active steps to dispel this view. A summit meeting held in December 1988 made the Rhodes Declaration. According to the declaration, Europe in 1992 would not transform itself into a fortress; instead, the European Union would contribute to greater liberalisation in international trade on the basis of GATT principles of reciprocal and mutually advantageous arrangements (*European Trends* 1989: 59).[35]

However, this declaration has done little to alleviate the concerns of many non-EU countries. Instead, many nations, including the Republic of Korea, are still concerned that Europe could be transformed into an economic fortress. That is, even though such concerns have weakened since 1988,[36] there is a fear that more protectionist measures could occur as a result of the 1992 EU process; clearly a number of worrying points still remain. Primarily, these concerns have to do with the fine print and subtle implications of regulatory policy. First is the notion of 'reciprocal and mutually advantageous arrangements'. Because access to one member country should mean access to the entire EU market, the Commission might insist on mirror-image reciprocity[37] in order to restrict the access of third parties.

Second, the European Commission is still as active as ever in taking up anti-dumping action against Japan. More recently, the Asian NICs have also become the target of the European Union's suspicion. In fact, anti-dumping action is a major factor of the EU in terms of external trade and, in recent years, the EU has aggressively developed anti-dumping action as a protective device, especially against imports from Japan and the Far East (Hindley 1988: 445; Hindley 1990: 4; Norall 1986: 109).

The rules of origin determine whether a particular batch of a product is subject to an anti-dumping duty or not (Hindley 1990: 11). In the EU, Council Regulation 802/68 controls determination of origin. Article 5 of that regulation states that:

> A product in the production of which two or more countries were concerned shall be regarded as originating in the country in which the last substantial process or operation that is economically justified was performed, having been carried out in an undertaking equipped for that purpose and resulting in the manufacture of a new product or representing an important stage of manufacture.
>
> (ibid.: 10)

Article 5, however, certainly has some room for debate in specific cases. There should be a more precise definition of the term 'substantial'. In fact, the European Union showed the inconsistency of its position with Regulation 802/68 by interpreting the definition of the term 'substantial' differently in dealing with the case of Brother typewriters in Taiwan[38] (ibid.).

In addition, the EU adopted in 1987 the so-called 'screwdriver-plant regulation' (Article 13(10) of the EU anti-dumping regulations), because the European Union thought that the exporters from the non-EU countries succeeded in 'circumventing' the anti-dumping duty in many cases (Commission of the European Communities 1988b). A foreign firm could establish EU origin merely by assembling in the EU, in a screwdriver plant, for example, components manufactured in another country.

Article 13(10) merely requires the investment and purchases necessary to avoid the duties it authorises to be undertaken outside the country of the alleged dumper. 'Under Article 13(10), an anti-dumping duty imposed on a final product could be extended to parts of the product shipped from the country of alleged dumping for assembly in the EU without separate investigation of the conditions of sale of the parts' (Hindley 1990: 4).

The EU also provided non-EU countries with two ways to escape the anti-dumping action through Article 13(10)(a). First, the non-EU manufacturers had to ensure that at least 40 per cent of the value of parts or materials for their assembly operations in the Union were produced outside the country of operation of the manufacturer. A second escape route was that the duties on components authorised by Article 13(10) could be applied only if 'the assembly or production operation was started or substantially increased after the opening of the anti-dumping investigation' (ibid.: 5). However, Article 13(10) also had some problems. Even though the article provided two ways to avoid the anti-dumping action, the entire article is open to subjective interpretation depending on each country's position. Because of this, non-EU countries could think that the two ways to avoid the anti-dumping action supplied by article 13(10) were not effective or reliable.

The confusion surrounding Article 13(10) can be seen from the following quotes. 'A GATT panel also had concluded that a requirement of written undertaking to purchase inputs in the host country was contrary to provision of the GATT' (Hayes 1990: iv). However, according to the *Financial Times* (8 January 1990), 'after months of dithering, the European Community finally acknowledged to its Uruguay Round partners that local content rules were an inadmissible barrier to foreign investment.' Even though the rules were not effective or even very clear, the fact that the European Union made such rules increased non-EU countries' fears about European integration.

The case of Brother typewriters and the EU's adoption of the 'screwdriver-plant regulation' make non-EU countries think that the European Union is creating 'strong' rules of origin. Strong rules of origin are rules with which it is expensive to comply. From the standpoint of any individual country, the arguments for 'strong' rules of origin are essentially the same as the arguments for high protection.

A Korean exporter, who thinks that he might in the future be hit by EU anti-dumping action, has an incentive to undertake current investment in the EU in order to avoid the worst effects of EU anti-dumping action. For these reasons, the 1992 process seems to be instilling fear in non-EU countries, and the EU is very important for Korean firms as an area for investment.

In addition, as of the end of 1995, the sheer size of the SEM was by itself very attractive to non-EU countries, since, with a combined population of 334 million, it exceeded that of both the US (263 million) and Japan (125 million).[39] It also had a GDP of US$ 7,082.7 billion in 1995, which matched that of the US (US$ 7,237.5 billion) and exceeded that of Japan (US$ 5,135.2 billion). Another attraction is that the birth of the SEM is expected to stimulate a spurt of economic activity, due to the removal of non-tariff and production barriers, scale economies and intensified competition.

Investment in the EU offers non-EU countries the gateway to market access or the opportunity for the direct investment in Eastern Europe because of the geographical and cultural proximity and the availability of market information (Lee 1990: 3). The now greater independence of the Eastern European countries, together with the reunification of East and West Germany, creates a rare market opportunity because of the high levels of demand for consumer goods due to inefficient resource allocation under the socialist system (ibid.). There are three ways in which resources have been misallocated (CEPR 1990: 3–4). First, the emphasis of heavy industry as a means of achieving both self-sufficiency and rapid industrial growth, second, the agricultural policy in the socialist countries of Europe has been an utter failure, with dramatic negative consequences for productivity; third, the systematic neglect of infrastructure, such as roads and telecommunications, market institutions, such as banks and other financial intermediaries, and the need for a well-functioning labour market.

As Table 1.18 shows, the share of consumer goods, such as foodstuffs and light industry, was reduced in the gross industrial output of the socialist-

Table 1.18 The share of the key industries in the gross industrial output of the socialist countries of Europe (per cent)

Industry	*Year*	*Bulgaria*	*Czecho-slovakia*	*GDR*	*Hungary*	*Poland*	*Romania*	*Yugoslavia*
Light industry	1950	21.0	11.2	12.3	13.4	15.5	15.9	17.9
	1980	11.9	9.2	8.7	8.3	12.0	14.0	13.2
Food	1950	41.0	25.1	14.3	21.4	31.7	25.1	19.9
	1980	20.2	13.9	15.8	14.6	14.9	12.9	10.7

Source: *Economic Geography of the Socialist Countries of Europe* (Alisov and Valev 1985: 11).

Note: Light industry: textiles, clothing, leather, fur and footwear industries.

countries of Europe for the period from 1950 to 1980, as a cost for the excessive investment in heavy industry.

Adding to the attractions of the EU the fact that it is one of few big markets which Korean firms can enter in order to diversify their markets in the near future. Korean firms also think that the ex-Soviet Union and the People's Republic of China have great potential as markets on a long-term basis, particularly in terms of potential market size.[40] However, on a short-term basis, it is very difficult to think of them as big markets, given their internal political and economic conditions.

During the Seoul Asian games in 1986, Chinese fever began in Korea. As of the end of 1994, there were 836 projects with China.[41] However, since the Tiananmen Square incident in June of 1989, the political situation in China is perceived as unstable, as is the economy. For these reasons, Korean firms have been very cautious in trade relations with China, although Korean investments show a dramatic increase in China in the 1990s.

In the case of the ex-Soviet Union, although it is difficult to analyse the current situation because of the poor quality of ex-Soviet economic statistics, the existence of a deep economic slump throughout the ex-Soviet Union, with a high rate of inflation over recent years, is obvious (Hillyard 1992: 3).

> During the first nine months of 1991 – especially after the August coup, the USSR's economic performance experienced a clear decline: GNP, −12 per cent; industrial production, −6.4 per cent; consumption, −17 per cent; cereal production, −25 per cent. According to NATO experts, the bottom has not yet been reached. Former USSR imports fell 40 per cent in 1991, while exports fell 10 per cent.
> (*Atlantic News* 22 January 1992)

The political situation in the ex-Soviet Union has also been in turmoil since the beginning of 1991, especially after the August military coup. The collapse of the ex-Soviet Union, followed by the serious economic slump and political instability, makes Korean firms think that it is not likely to become a major market in the near future.[42]

1.3 Conclusion

After independence from Japan in 1945, Korea was faced with a very difficult period for about 15 years. The Korean War lasted for three years, from 1950 to 1953, and shattered the fledgling Korean economy. Since the early 1960s, however, the Korean government has achieved miraculous economic growth through a series of five-year Economic Development Plans, starting in 1962. During the period from 1962 to 1988, the average annual economic growth rate was about 8.7 per cent and the per capita GNP increased by 46 times at current prices. As of the end of 1989, according to GATT (Korea

Trade Promotion Corporation 1990), Korea had the thirteenth-largest import and export volumes in the world, making it a substantial economic power.

As a result of the export-oriented policy that the Korean government adopted in the early 1960s because of the relatively small size of the home market, the Korean economy is heavily dependent on external markets. The Korean economy has various difficulties because of external factors, such as vulnerability to world-wide recession and protectionism, and internal factors, such as the unbalanced trading relations concentrated in a few advanced countries, especially in the USA and Japan, and labour unrest in 1987 and 1988.

To overcome these difficulties, the Korean government needs to take appropriate action. Diversifying trading relations will be one of the best ways of solving the difficulties caused by external factors. In this context, the EU is clearly an important future market for Korea.

2

OVERVIEW OF KOREAN FDI

During the 1950s and 1960s, firms in the United Kingdom and the United States were by far the biggest investors in other countries. In the early 1970s German and Japanese firms emerged as new big foreign investors.[1] Since the mid-1980s, however, firms based in LDCs, and especially those based in the NICs, have also emerged as significant overseas investors by increasing their production in other LDCs and, in a few instances, in developed countries.[2] Overseas investment by firms from LDCs was US$ 6 billion in 1988, US$ 10 billion in 1989, US$ 10 billion in 1990, US$ 7 billion in 1991 and US$ 9 billion in 1992 (UN 1994: 12). Since the mid-1980s, manufacturing firms from developing countries and NICs have entered international markets in large and growing numbers. Korean firms are no exception.

The main purposes of this chapter, then, are to analyse the characteristics, patterns, recent trends and distribution of Korean FDI in general and in the EU in particular. To do this, this chapter has the following five parts. The first section is an overview; the second analyses the general trends of Korean FDI; next, the characteristics, patterns, recent trends and distribution of Korean FDI in the EU are analysed in comparison with Japanese and Taiwanese firms in the EU. In the fourth section, the general trends of FDI by Korean consumer electronics firms in the EU are analysed. The final section is a conclusion.

2.1 Overview of recent Korean FDI

In 1968, Korea's first foreign investment, of US$ 3 million, was made for a timber project in Indonesia by a Korean firm (P.T. Development Co. Ltd). Korea's overseas investment began in natural resources, developing raw material supplies and also in the wholesale trade sectors of conglomerate trade companies (*chaebols*). Later it extended to manufacturing and service sectors such as banking.

At the end of 1994, the total stock of Korea's net overseas investment[3] stood at US$ 7.65 billion for 4,161 projects, as Table 2.1 shows. Korean FDI was, however, still at an early stage compared to those of other developed

Table 2.1 Korea's net overseas investment (1968–94)

Year	*No. of projects (cases)*	*Value (US$ 1,000)*	*Average value of Investment (US$ 1,000)*
68–75	97	46,547	480
1976	15	6,300	420
1977	45	18,729	416
1978	73	39,544	542
1979	42	16,397	390
1980	15	12,795	853
1981	38	28,793	758
1982	31	97,511	3,146
1983	44	102,592	2,332
1984	33	48,183	1,460
1985	10	66,540	6,654
1986	33	161,184	4,884
1987	60	320,897	5,348
1988	144	164,096	1,140
1989	246	392,387	1,595
1990	317	813,185	2,565
1991	430	1,037,224	2,412
1992	465	1,133,683	2,438
1993	616	1,071,389	1,739
1994	1,407	2,070,816	1,472
Total	4,161	7,648,792	1,838

Source: *HaeWae TuJa TongKae YunBo* [Overseas Direct Investment Statistics Yearbook] (Bank of Korea 1995).

Notes: No. of projects = no. of total investment projects – no. of disinvestment projects. Amount = Amount of total investment projects – amount of disinvestment projects.

countries and Asian NICs. For example, in terms of the share of total overseas investment in the nation's GNP in 1994, Korea's share was only 2.0 per cent, while those of Japan and Taiwan were 9.8 per cent and 3.6 per cent, respectively, as Table 2.2 illustrates.

As illustrated in Table 2.2, as of the end of 1994, the cumulative value of Korea's outward FDI was about US$ 7,649 million, which was about 86 per cent of the Taiwanese and only 1.6 per cent of the Japanese levels. This accumulated value of Korea's outward FDI was 8.0 per cent of its exports in 1994, which was also lower than that of Taiwan and Japan, which were 9.6 per cent and 116.8 per cent, respectively. The number and flow value of Korea's overseas investment was very meagre before 1986, as Table 2.1 shows. Korean firms did not make FDI very actively because of the shortage of capital, limited information, lack of competitiveness of Korean firms in the international markets and an absence of incentive measures from the government.

Table 2.2 Korean FDI compared with Japan and Taiwan (1994) (US$ million)

Accumulated value of FDI by 94 (Approval basis)			*GNP of 94 (C)*	*Export of 94 (D)*	*Import of 94 (E)*	*A/C (%)*	*A/D (%)*	*B/E (%)*	*B/A (%)*
	Outward (A)	*Inward (B)*							
Korea	7,649[F]	12,525[F]	378,000	96,013	102,348	2.0	8.0	12.2	163.7
Japan	463,606	34,088	4,725,759	397,000	275,200	9.8	116.8	12.4	7.3
Taiwan	8,897	19,335[G]	244,316	93,049	85,349	3.6	9.6	19.5	217.3

Source: *Major Statistics of Korean Economy* (National Statistical Office, Republic of Korea 1996), *Direction of Trade Statistics* (IMF 1995); *Statistical Yearbook* (Directorate-General of Budget, Accounting and Statistics, Republic of China 1996); *Taiwan Statistical Data Book* (Council for Economic Planning and Development, Republic of China 1996); *Analysis of Finance: Japan* (JETRO 1995).

Notes: C: GNP is at current prices. D: Exports on free an board (FOB) status E: Imports are based on CIF. F: Net permitted investment = total permitted investment − Invalidated investment. G: This figure includes private foreign and overseas Chinese investment.

Korean FDI had increased steadily since 1968, even though it was still at an early stage, as explained above. Between 1968 and 1982, Korea's overseas investment increased very slowly. The year 1978 was an exception. In 1978 Korean FDI increased substantially, even though from a meagre base. This large increase was due primarily to a sharp rise in investment in construction projects in the Middle East (Sakong 1993: 150).

During two years, 1982–3, after the second oil shock, Korea's overseas investment recorded a big increase due to the dramatic increase of investment in the mining sector. Since 1982, there has been large-scale foreign direct investment in resource-development projects, especially coal mining projects. The number of such projects was not large, but the value of investment was substantial because each project required huge amounts of capital (ibid.: 151). The large-scale FDI in coalmining projects increased the average value of Korea's overseas investment projects. In the 1970s, the average value of investment projects was about US$ 400,000. The average value of investment projects for the period from 1968 to 1994 was about US$ 1.8 million, as Table 2.1 illustrates. The big increase of FDI in the mining sector reflected the difficulties experienced by the Korean government and business because of the two oil shocks. In more detail, in 1982, Korea's net overseas investment recorded a 246 per cent increase (US$ 28 million in 1981 to US$ 97 million in 1982) over the previous year. The value of Korean net overseas investment projects in 1983 also increased by 5 per cent. However, there was a short downturn in 1984 due to the decrease in natural resources development investment.[4] However, the flow of net Korean overseas investment continued to rise through 1985–6, and peaked in 1987, when it recorded US$ 320 million.

The upsurge of Korean overseas investment had begun in 1986 when the economy experienced a trade surplus for the first time in its history.[5] Owing to the trade surplus, the Korean government decided to lift a substantial proportion of its foreign exchange controls and began encouraging FDI by Korean firms and individuals.[6] During the period from 1986 to 1988, Korea experienced unprecedented real economic growth. The GNP growth rate in real terms recorded an average gain of more than 12 per cent per annum.[7] This trend in GNP was possible owing to a combination of such favourable factors as low interest rates, low oil prices, the weakness of the won and the low rate of wage increases, as well as an industrial infrastructure which could readily take advantage of such factors. As a result, the Korean economy produced competitive exports. Because of the sharp increase in its current account surplus,[8] Korea's outstanding net foreign debt, as of the end of 1989, was reduced to US$ 29.4 billion from US$ 46.7 billion in 1985.

For the last 20 years, Korean firms have pursued an export-market-oriented development strategy. Korea's poor resource-endowment has called for economic growth based on manufacturing activities. Korea has a limited domestic market that could not provide opportunities for further import substitution. Thus, industrial expansion had to be sought through exporting. FDI was the next step after the export-oriented strategy of the 1970–85 period to supply foreign markets for local production. In fact, recent changes in the economic environment surrounding the Korean economy induced Korean firms to take overseas operations seriously. Such external factors as the stiff trade barriers of the developed countries and the birth of strong regionalism in the world trade system, and such internal factors as the Korean won appreciation[9] and wage surge[10] worked as positive forces. The appearance of multilateralism and growth of regionalism in the world trading system encouraged Korean firms to establish production facilities in the existing and potential foreign markets in order to reduce the vulnerability from any renewed protectionist measures.[11] Such development in the world trading system particularly encouraged more Korean investment in the more technologically advanced manufacturing sectors. Knowledge-oriented FDI contributed a way of overcoming the barriers in the way of technology transfer imposed by developed countries and of acquiring high technologies from developed countries.

Second, with the appreciation of the Korean won induced by the current account surpluses and the rapid domestic wage increases, the labour-intensive industries had to compete by means of improved quality rather than lower prices, or by relocating their facilities to relatively labour-abundant countries.

Finally, FDI was recommended by both the Korean government and firms because it was expected to ease trade frictions and control liquidity accompanying external surpluses. With the increasing need for continued adjustment of international payment imbalances, accumulation of external surpluses began to provoke protectionist sentiment against Korean products in foreign countries, especially those which had trade deficits with Korea (such as the USA in the mid-1980s). This made it difficult for Korean firms to export their products to foreign countries. In sum, a new purpose for FDI had arisen – as a means of globalising management, marketing and production.

Due to the changing environment, Korean firms had to adopt a so-called, 'globalisation strategy,' in order to at least maintain competitiveness. The 10 biggest enterprises in Korea (Hyundai, Lucky GoldStar, Samsung, Daewoo, Sunkyung, Hankuk Hwayak, Ssangyong, Tusan, Kolon and Hanjin) showed this change in their 1986 New Year's messages. They announced in their messages that they intended increasingly to expand their overseas activities and investments (*Dong-A Ilbo* (newspaper) 7 January 1986). My survey, in fact, revealed that the main rationale cited by Korean consumer electronic firms for FDI in the EU was to implement a 'globalisation strategy'.[12] This meant Korean firms changing their globalisation strategy from an export-market-oriented to foreign local-production-oriented strategy, despite risks involved in the latter.

Although the trade balance has worsened since 1989, both the number and absolute value of FDI projects of Korean firms have sharply increased, as illustrated in Table 2.1. The surge of Korean FDI is the result of a new strategy of Korean firms' globalisation.

Korea's overseas investment can be characterised by industry, by region, by pattern, by investment size and by investing firms' size. By industry, as of the end of 1994, both the number and value of foreign investment projects were the highest in manufacturing (textiles and clothing, petroleum and fabricated metals), trade and mining, as illustrated in Table 2.3. This is in contrast to previous overseas investment, which was largely oriented to the development

Table 2.3 Korea's net overseas investment by industry (as of 31 December 1994)

	No. of investment projects (%)	*Value (US$ million) (%)*
Manufacturing	2,652 (63.7)	4,191 (56.1)
Fishing	122 (2.9)	105 (1.4)
Forestry	18 (0.4)	79 (1.1)
Mining	50 (1.2)	678 (9.1)
Real estate	30 (0.7)	153 (2.0)
Trading	807 (19.4)	1,751 (23.4)
Construction	95 (2.3)	138 (1.8)
Transportation	65 (1.6)	41 (0.5)
Others	322 (7.7)	514 (6.9)
Total	4,161 (100.0)	7,469 (100.0)

Source: *HaeWae TuJa TongKae YunBo* [Overseas Direct Investment Statistics Yearbook] (Bank of Korea 1995).

Note: Because of rounding, figures may not add to the totals shown.

of natural resources. More recently, attention has focused on manufacturing and trade in order to overcome trade barriers in the developed countries, the appreciation of the Korean won and rising labour costs. As of the end of 1994, the manufacturing sector accounted for 63.7 per cent of the total number of overseas investment projects, while the trading sector accounted for 19.4 per cent. As of the end of 1994, the manufacturing sector accounted for 56.1 per cent of the total value of foreign investment projects, followed by 23.4 per cent in the trading sector. The average value of FDI projects in the manufacturing sector was much less than that in the trading sector.

Geographically, investment has been concentrated in North America, especially the USA, which has been the major market for exports and in Southeast Asia, with its abundant manpower and natural resources. Recently, however, Korean overseas investment has begun to diversify into other areas such as Europe, South and Central America and Oceania. Despite the efforts to diversify

Table 2.4 Korea's overseas investment in total and in manufacturing by region (as of 31 December 1994)

	No. of projects (1) (%)	*No. of projects (2) (%)*	*2/1*	*Amount (US$ million) (3) (%)*	*Amount (US$ million) (4) (%)*	*4/3*
Asia (SE)	2845 (68.4)	2201 (83.0)	77.4%	3,202 (41.9)	2,284 (54.5)	71.3%
Oceania	111 (2.7)	33 (1.2)	29.7%	221 (2.9)	22 (0.5)	10.0%
Africa	45 (1.1)	20 (0.8)	44.4%	230 (3.0)	83 (2.0)	36.1%
Europe	245 (5.9)	82 (3.1)	33.5%	981 (12.8)	505 (12.0)	51.5%
L. America	218 (5.2)	121 (4.6)	55.5%	238 (3.1)	125 (3.0)	52.5%
N. America	662 (15.9)	184 (6.9)	27.8%	2,704 (35.4)	1,154 (27.5)	42.7%
Mid-East	35 (0.8)	11 (0.4)	31.4%	73 (1.0)	18 (0.4)	24.7%
Total	4,161 (100)	2,652 (100)	63.7%	7,649 (100)	4,191 (100)	54.8%

Source: *HaeWae TuJa TongKae YunBo* [Overseas Direct Investment Statistics Yearbook] (Bank of Korea 1995).

Notes:
1. Total number of FDI cases.
2: Number of FDI projects in manufacturing sector.
3: Total amount of FDI projects.
4: Amount of FDI projects in manufacturing sector.
Because of rounding, figures may not add to the totals shown.

its investment, as of the end of 1994, Southeast Asia and North America still accounted for 84.3 per cent of the total number of overseas investment projects and 77.3 per cent of the total stock value of overseas investment projects, as Table 2.4 shows.

The investment in manufacturing took 83 per cent of the total accumulated number of FDI projects and 54.5 per cent of the total accumulated value of FDI projects in Southeast Asia. The manufacturing sector took only 6.9 per cent of the total accumulated number of investment projects in North America and 27.5 per cent of the total accumulated value of investment projects in North America, respectively.

In terms of investment size, 76.3 per cent of the total accumulated number of FDI projects, as of the end of 1994, were less than US$ 1 million, while only 13 per cent were more than US$ 2 million, as illustrated in Table 2.5.

Finally, Koreans investing overseas show a strong preference for wholly owned investment rather than joint ventures, as Table 2.6 illustrates. As of the end of 1994, wholly owned investments took the highest proportion at 50.9 per cent of the total cumulative number of FDI projects, while less than 20 per cent owned investments took only 2.2 per cent.

Table 2.5 Korea's investment position by investment scale (as of 31 December 1994)

Total	*<US$ 1 million*	*>US$ 1 million <US$ 2 million*	*>US$ 2 million <US$ 5 million*	*>US$ 5 million <US$ 10 million*	*>US$ 10 million*
4,161 (100%)	3,174 (76.3%)	445 (10.7%)	315 (7.6%)	114 (2.7%)	113 (2.7%)

Source: *HaeWae TuJa TongKae YunBo* [Overseas Direct Investment Statistics Yearbook] (Bank of Korea 1995).

Table 2.6 Korea's overseas investment position by ownership share (as of 31 December 1994)

Total	*< 20%*	*20%–50%*	*50%*	*50%–100%*	*100%*
4,161 (100%)	93 (2.2%)	705 (16.9%)	392 (9.4%)	854 (20.5%)	2,117 (50.9%)

Source: *HaeWae TuJa TongKae YunBo* [Overseas Direct Investment Statistics Yearbook] (Bank of Korea 1995).

Note: Because of rounding, figures may not add to the totals shown.

2.2 Korean FDI in the EU

FDI of Korean firms in the EU has increased rapidly since the mid-1980s, as illustrated in Table 2.7. This is due to changes in the international market

Table 2.7 Korean FDI in the EU (1986–94)

(Unit: thousands of US$, cases)

		1986	*1988*	*1990*	*1992*	*1994*
EU (1)	Accumulated value of FDI	17,544	65,053	160,600	279,946	852,383
	Accumulated cases of FDI	47	69	104	123	165
Rest of the World	Accumulated value of FDI	975,488	1,610,280	3,538,235	4,226,641	6,796,409
	Accumulated cases of FDI	459	724	1,471	2,015	3,996

Sources: *Hae Wae TuJa TongKae YunBo* [Overseas Direct Investment Statistics Yearbook] (Bank of Korea 1991, 1992, 1995), *HaeWae TuJa HyunJi BupIn HyunWhang* [Current Status of Korean Affiliates Abroad] (Bank of Korea 1996).

Note: 1: Calculated from *HaeWae TuJa HyunJi BupIn HyunWhang* [Current Status of Korean Affiliates Abroad] (Bank of Korea 1996).

situation, such as world-wide protectionism and regionalism. In fact, it is very important for Korean firms to increase their direct investment in the EU in order to diversify export markets and circumvent trade disputes, especially with such developed countries as the USA and Japan. However, the stock of Korean FDI in the EU is still meagre compared to those in North America and Southeast Asia. In fact, Korea's inward and outward FDI in the EU represents a small percentage both in the total cumulative number and amount of Korean FDI projects. As of the end of 1994, FDI in the EU took only 4 per cent and 11.1 per cent in the total cumulative number and amount of outward Korean FDI, respectively. EU's FDI in Korea also took only 10.4 per cent and 20.8 per cent in the total cumulative number and amount of inward Korean FDI, respectively.[13]

It is interesting to note, however, that there are differences between Korea's direct investment in the EU and that in other regions. First, the average value of Korean direct investments in the EU grew faster in the period than that elsewhere. On the whole, the average value of overseas investment by Korean firms increased quite rapidly. It was US$ 1.36 million as of 1986, US$ 1.80 million as of 1987, US$ 1.66 million as of 1988, US$ 1.64 million as of 1989 and US$ 1.9 million as of 1990. The average value of each FDI project in the EU was US$ 0.38 million as of 1986, US$ 0.45 million as of 1987, US$ 0.64 million as of 1988, US$ 0.45 million as of 1989, US$ 1.5 million as of 1990 and US$ 5.2 million as of 1994 (see Table 2.8).

Second, as Table 2.9 demonstrates, as of the end of 1994, the value of Korean FDI in the manufacturing sector took 36 per cent of the total value of FDI projects in the EU member states. This share was lower than in Southeast Asia (71 per cent), Latin America (53 per cent) and North America (43 per cent),

Table 2.8 Korea's direct investment position by investment scale (as of 31 December 1994)

Area		*< US$ 1 million*	*US 1 million–5 million*	*Over US$ 5 million*	*Total*	*Average size of project*
Asia (SE)	No. of projects	2,284	478	83	2,845	1,125
	Amount	612,966	1,025,067	772,244	3,201,820	
Mid–East	No. of projects	25	8	2	35	2,098
	Amount	9,915	11,781	51,738	67,873	
N. America	No. of projects	450	142	70	662	4,084
	Amount	145,746	340,249	2,217,850	2,703,845	
L. America	No. of projects	156	55	7	218	1,090
	Amount	63,674	120,131	53,707	237,512	
Europe–(1)	No. of projects	145	52	48	245	4,005
	Amount	40,469	142,860	797,971	981,300	
EU (12 states)	No. of projects	82	40	43	165	5,166
	Amount	21,783	99,722	730,878	852,383	
Africa	No. of projects	31	7	7	45	5,109
	Amount	6,894	13,822	272,202	229,918	
Oceania	No. of projects	83	18	10	81	1,982
	Amount	28,141	38,615	154,207	220,963	
Total	No. of projects	3,174	760	227	4,161	1,838
	Amount	906,656	1,692,525	5,048,462	7,648,792	

Sources: *Hae Wae TuJa TongKae YunBo* [Overseas Direct Investment Statistics Yearbook] (Bank of Korea 1995); *Hae Wae TuJa HyunJi BupIn HyunWhang* [Current Status of Korean Affiliates Abroad] (Bank of Korea 1996).

Note: 1: Projects and amount including those in the EU.

although it was higher than in the Middle East and Oceania. In the number of Korean FDI projects in the EU, as of the end of 1994, the share of the manufacturing sector (30 per cent) was smaller than that in all other regions except North America which had also 30 per cent.

Korean FDI has different investment motivations depending on regions. Korean FDI in the EU and North America was motivated by the need to obtain further market penetration through establishing subsidiaries for manufacturing and wholesale trade, while that in Southeast Asia, the Middle East and Oceania was aimed at securing the supply of mining and other raw materials, including fishing and wood products. This is in sharp contrast with

Table 2.9 Korea's direct investment position abroad (as of 31 December 1994)

Area		*Mining*	*Manufacturing*	*Wholesale trade*	*Others (2)*	*Total*
Asia (SE)	A	16	2,201	312	316	2,845
		(1%)	(77%)	(11%)	(11%)	(100%)
	B	341,642	2,284,283	235,782	340,113	3,201,820
		(11%)	(71%)	(7%)	(11%)	(100%)
Mid–East	A	2	11	2	20	35
		(6%)	(31%)	(6%)	(57%)	(100%)
	B	39,657	18,404	370	15,003	73,434
		(54%)	(25%)	(1%)	(20%)	(100%)
N. America	A	10	184	315	153	662
		(2%)	(28%)	(48%)	(23%)	(100%)
	B	110,994	1,153,762	1,062,321	376,768	2,703,845
		(4%)	(43%)	(39%)	(14%)	(100%)
L. America	A	2	121	21	74	218
		(1%)	(56%)	(10%)	(34%)	(100%)
	B	13,861	125,256	21,816	76,579	237,512
		(6%)	(53%)	(9%)	(32%)	(100%)
Europe	A	3	82	130	30	245
		(1%)	(34%)	(53%)	(12%)	(100%)
	B	45,454	504,798	408,306	22,742	981,300
		(5%)	(51%)	(42%)	(2%)	(100%)
EU	A	0	46	100	19	165
		(0%)	(28%)	(61%)	(12%)	(100%)
	B	0	303,283	393,052	156,048	852,383
		(0%)	(30%)	(46%)	(18%)	(100%)
Africa	A	3	20	7	15	45
		(7%)	(44%)	(16%)	(33%)	(100%)
	B	32,491	82,857	8,246	106,324	229,918
		(14%)	(36%)	(4%)	(46%)	(100%)
Oceania	A	14	33	20	44	111
		(13%)	(30%)	(18%)	(40%)	(100%)
	B	93,962	21,741	13,721	91,539	220,963
		(43%)	(10%)	(6%)	(41%)	(100%)
Total	A	50	2,652	807	652	4,161
		(1%)	(64%)	(19%)	(16%)	(100%)
	B	678,061	4,191,101	1,750,562	1,029,068	7,648,792
		(9%)	(55%)	(23%)	(13%)	(100%)

Sources: *HaeWae TuJa HyunJi BupIn HyunWhang* [Current Status of Korean Affiliates Abroad] (Bank of Korea 1996); *HaeWae TuJa TongKae YunBo* [Overseas Direct Investment Statistics Yearbook] (Bank of Korea 1995).

Notes: 1– A: is no. of projects. B: is amount of project in US$ 1,000. Others include fishery, forestry, construction, transportation, real estate, etc.

Because of rounding, figures may not add to the totals shown.

Japan, whose direct investment in the EU is heavily directed towards the banking and real estate sectors. In fact, Japanese investment in Europe[14] over the same period was dominated by banking and insurance companies (Thomsen and Nicolaides 1992: 8).

During the period from 1951 to 1989, cumulative Japanese investment in banking and insurance sectors took 47 per cent out of the total cumulative Japanese investment in Europe. There are several reasons for the overseas expansion of banking and insurance sectors. According to Kitamura (1991: 86), first, the increasing need from Japanese firms for diversified financial services in raising capital and fund management contributed to the overseas expansion of these sectors. The changes in the 1980s (such as financial deregulation and liberalisation in major countries and technological innovation in computerisation and telecommunications in the financial sector) certainly enabled Japanese financial institutions to do more efficient financial transactions in foreign countries. Second, financial assets which had been rapidly accumulated provided wider international business scope for Japanese financial institutions. The stable economic performance in the 1980s and the appreciation of the yen since 1985[15] also contributed to the rapid overseas expansion of the financial sector by strengthening the base of the financial institutions in the international markets.

As of the end of 1994, the EU was the only region where the proportion of investment to wholesale trade was greater than that of manufacturing investment in the total value of FDI projects. The proportion of investment value in wholesale trade, as of the end of 1994, was 46 per cent, while that in manufacturing was 36 per cent in the EU, as illustrated in Table 2.9.

The investment in manufacturing as a proportion of the total number of FDI projects showed a slightly different pattern in the EU, as of the end of 1994. The EU was not the only region where the proportion of investment to wholesale trade in the total number of FDI projects was greater than that of investment to manufacturing, as Table 2.9 shows. The proportion of investment to wholesale trade in the total number of FDI projects in North America was also greater than that of investment to manufacturing. The proportion of investment to wholesale trade in North America, as of the end of 1994, was 48 per cent, while that of investment to manufacturing was 28 per cent. As of the end of 1994, the average proportion of manufacturing investment and wholesale trade investment out of the total number of FDI projects worldwide was 64 per cent and 19 per cent, respectively, as Table 2.9 illustrates.

The fourth distinctive feature in Korea's FDI in the EU is the ownership share. As of the end of 1994, Korea's direct investment in the EU was more inclined to focus on the majority-owned or wholly owned operation (86.4 per cent) than in other regions except Oceania (91.9 per cent) and North America (86.4 per cent), as illustrated in Table 2.10. This ownership pattern has arisen since most Korean subsidiaries are sales representatives of the integrated trading companies at home.

Table 2.10 Korea's direct investment position by ownership share (as of 31 December 1994) (projects, percentage for total no. of projects)

Area	*0%–19%*	*20%–49%*	*50%–99%*	*100%*	*Total*
Asia (SE)	41 (1.4%)	555 (19.5%)	985 (34.6%)	1,264 (44.4%)	2,845 (100%)
Mid-East	2 (5.7%)	22 (62.9%)	9 (25.7%)	2 (5.7%)	35 (100%)
N. America	35 (5.3%)	55 (8.3%)	101 (15.3%)	471 (71.1%)	662 (100%)
L. America	4 (1.8%)	29 (13.3%)	40 (18.3%)	145 (66.5%)	218 (100%)
Europe (A)	9 (3.7%)	28 (11.4%)	69 (28.2%)	139 (56.7%)	245 (100%)
(EU)	5 (3.2%)	9 (5.8%)	22 (14.1%)	129 (72.3%)	165 (100%)
Africa	1 (2.2%)	8 (17.8%)	17 (37.8%)	19 (42.2%)	45 (100%)
Oceania	1 (0%)	8 (7.2%)	25 (22.5%)	77 (69.4%)	111 (100%)
Total	93 (2.2%)	705 (16.9%)	1,246 (29.9%)	2,117 (50.9%)	4,161 (100%)

Sources: *HaeWae TuJa HyunJi BupIn HyunWhang* [Current Status of Korean Affiliates Abroad] (Bank of Korea 1996); *HaeWae TuJa TongKae YunBo* [Overseas Direct Investment Statistics Yearbook] (Bank of Korea 1995).

Notes: Because of rounding, figures may not add to the totals shown. A: no. of projects and per cent, including those in the EU.

In addition, of the projects of Korea's direct investment in the EU, North America and Oceania, about 70 per cent were wholly owned, whereas only about 50 per cent of the projects of Korea's direct investment in other regions, as of the end of 1994, were wholly owned. This feature seems to be inconsistent with the prediction of direct investment theory. Direct investment theory posits that firms without technological superiority over their local competitors tend to prefer mergers and acquisitions (M&A) or joint ventures to new or wholly owned subsidiaries.

Fifth, major sectors of manufacturing that Korean firms established in the EU were assembly lines of machinery and related products such as video equipment and accessories, CTVs, computer components, home appliances, and motor vehicles and equipment.

2.2.1 FDI of Korean electronics firms in the EU

As discussed earlier, Korean FDI in the EU has increased rapidly both in its value and the number of projects since the mid-1980s. At that time, the FDI of consumer electronics firms took the largest portion of Korean manufacturing FDI in the EU both in its value and the number of projects. As of the end of 1994, as illustrated in the Table 2.11, the value of FDI of Korean consumer electronics firms in the EU was U.S.$ 256.8 million, which was 85 per cent of the total value of Korean FDI in the manufacturing sector in the EU (US$ 303.3 million). The number of FDI projects of Korean consumer electronics firms in the EU was about 54 per cent of the total number of FDI projects of Korean FDI in the manufacturing sector in the EU.

FDI of Korean electronics firms in the EU also had several characteristics. As of the end of 1994, Korean firms had 165 cases of direct investment in the EU, which amounted to about US$ 852 million, as Table 2.9 demonstrates. Out of 165 projects of Korean FDI in the EU, 100 projects were in the wholesale trade sector and 46 projects were in the manufacturing sector.

Table 2.11 FDI of Korean electronics firms in the EU (1982 and 1986–94) (cases, US$ million)

	1982	*1986*	*1987*	*1988*	*1989*	*1990*	*1991*	*1992*	*1993*	*1994*	*Total*
No. of projects	1	1	1	3	4	4	4	2	3	2	25
Value of FDI	6.3	39.3	22.5	19	7.7	26	9.1	25.1	64.2	37.6	256.8

Source: *Hae Wae Tuja HyunJi BupIn Hyunwhang* [Current Status of Korean Affiliates Abroad] (Bank of Korea 1996).

Table 2.12 Korean electronics firms in the EU (as of 31 December 1994)

	Approval date	*Ownership scale*	*Approved amount of investment (US$ 1,000)*	*Implemented amount of investment (US$ 1,000)*
FRANCE				
Daewoo Electronics	88.8.17	65%	1,879	1,879
Haitai Euro S.A.	89.9.27	51%	680	680
Daewoo Electronics Manufacturing S.A.	92.4.25	100%	29,353	21,353
Daewoo Orion S.A.	93.7.15	100%	60,000	60,000
ITALY				
LG Electrodomestici S.P.A.	90.1.25	40%	3,701	3,701
Samsung Electronics	89.9.25	20%	140	140
GERMANY				
Saejon	91.2.11	20%	12	12
Goldstar	86.9.4	100%	39,258	39,258
Samsung Elektronische Bauelemente GmbH.	92.12.10	100%	3,741	3,741
Samsung Corning Deutschland GmbH.	94.2.15	100%	33,670	9,670
UK				
Samsung Electronics UK Ltd	84.3.7	100%	34,952	34,952
Goldstar	88.10.13	100%	12,090	12,090
Goldstar Elec. Wire	91.8.5	26.04%	337	337
Daewoo Electronics UK Ltd	88.11.5	100%	5,000	5,000

Table 2.12 (contd)

	Approval date	*Ownership scale*	*Approved amount of investment (US$ 1,000)*	*Implemented amount of Investment (US$ 1,000)*
SKM	91.5.30	98%	2,006	2,006
Kolon	91.9.13	100%	6,786	6,786
Ssangyong/Trigem Computers UK Ltd	89.9.8	100%	2,010	2,010
Inkel	90.7.10	100%	4,082	3,102
CCA Electronic Ltd	93.3.18	100%	3,000	3,000
Sammi Sound Technology UK Ltd	93.9.22	100%	1,213	1,213
PORTUGAL				
Samsung Electronica Portuguesa S.A.	82.7.5	100%	6,286	6,286
Samsung Portugal Produtos Electro-mecanicos S.A.	90.5.22	100%	18,140	18,140
Tisep-Electronica, Portugal, LDA	94.3.22	38%	31,843	27,920
(Samsung Electronics)				
IRELAND				
Kony	90.6.12	50%	1,000	1,000
Saehan	87.4.21	100%	22,500	22,500
SPAIN				
Samsung Electronics	89.1.17	100%	4,838	4,838
TOTAL	—	—	328,517	291,614

The 46 European units of Korean manufacturers can be broken down by industrial line into 26 electronics and electrical machinery and parts makers, six general machinery makers, three chemical firms, three non-metal and basic metal makers, two leather goods makers, two textiles and clothing makers, two paper and printing firms and two food makers. Direct investment of Korean electronics and electrical makers in the EU shows the following features. First, by far the largest number of electronic and electrical machinery and parts makers with Korean capital affiliation existed in the United Kingdom, which had ten such investments, followed by Germany and France with four each, Portugal with three, and Italy and Ireland with two each.

The location distribution of Japanese manufacturing firms in the EU showed a very similar tendency to the Korean case, as Table 2.13 shows. According to the results of a survey[16] of host countries, as of the end of 1995, by far the largest number of Japanese capital-affiliated manufacturers

Table 2.13 Japanese manufacturing enterprises in the EU classified by country (1985–95)

Country	*1985*	*1986*	*1987*	*1988*	*1989*	*1990*	*1991*	*1992*	*1993*	*1994*	*1995*
UK	46	55	77	98	127	158	180	191	200	204	217
France	32	40	52	61	73	83	92	96	97	103	109
Germany	41	48	55	60	75	83	91	99	102	102	105
Netherlands	17	18	21	24	29	29	37	42	44	45	48
Belgium	18	21	22	25	29	36	38	39	41	42	42
Luxembourg					1	1	2	2	2	2	2
Ireland	9	11	12	14	17	23	25	26	30	30	33
Spain	23	26	31	36	44	48	53	55	56	59	60
Italy	12	15	20	27	33	36	41	42	43	45	49
Denmark	1	1	1	1	2	2	2	2	2	2	2
Portugal	4	5	5	7	9	11	11	11	12	13	14
Greece	3	3	3	3	3	3	3	3	3	3	3
Total	206	243	299	356	442	513	575	608	632	650	684
Yearly change in number		37	56	57	86	71	62	43	24	8	34
Change over the year (%)		20	23	19	24.2	16.1	12.1	5.7	3.9	2.8	5.2

Source: Calculated from *Twelfth Survey of European Operations of Japanese Companies in the Manufacturing Sector* (JETRO 1996).

Note: Including manufacturing bases, but excluding independent R&D facilities.

were in the United Kingdom, which had 217 (29.8 per cent of the total), followed by France with 109 (15 per cent), Germany with 105 (14.4 per cent), Spain with 60 (8.3 per cent) and Italy with 49 (6.7 per cent).

As of the end of 1995, 727 Japanese-affiliated manufacturers were present in Europe. The 727 European units of Japanese manufacturers can be broken down by industrial line into 170 electronic and electrical machinery and parts makers (23.4 per cent of the total number of projects), 130 chemical firms (17.9 per cent), 94 general machinery makers (12.9 per cent), 74 transport equipment and parts manufacturers (10.2 per cent), and 39 precision instruments makers (5.4 per cent). These five industries together accounted for 69.7 per cent of the total. Parts makers total 114, of which 59 firms were manufacturing electronic and electrical machinery parts and 55 were transport equipment parts manufacturers.

In addition, the largest number of electronic and electrical machinery and parts makers with Japanese capital affiliation existed in the United Kingdom, which had 67, followed by Germany with 32 and France with 26, as of the end of 1995 (JETRO 1996). These three countries represent 59 per cent of the European markets where Japanese companies have made inroads.

Second, direct investment of Korean electronics firms in the EU also inclined to be majority-owned or wholly owned operations. Out of 26 direct investments in the sector, 17 projects were wholly owned operations and four projects were majority-owned operations with more than 50 per cent ownership share, as illustrated in Table 2.12. It is noteworthy that nine out of ten projects of direct investment of Korean electronics firms in the UK were majority-owned operations, while neither of the direct investment projects in Italy were majority-owned, as shown in Table 2.11.

Third, the average value of direct investment projects of Korean electronics firms in the EU, as of the end of 1994, was about US$ 10.3 millions. In more detail, 26 investment projects were made in the EU as of the end of 1994. Among these, there were 12 large investments with more than US$ 5 millions (46 per cent), and 10 projects of investments (39 per cent) were middle-sized ones that were in the range from more than US$ 1 million to US$ 5 millions. The four remaining investment projects (15 per cent) were small-sized ones that were less than US$ 1 million.

2.3 Conclusion

The Korean government and firms are now reassessing their strategies concerning investment for future economic development against a trend of growing protectionism and regionalism. In Korea, economic instability resulted from internal problems such as rising wages and appreciation of the won. Moreover, the substantial balance of payments surplus recorded from 1986 to 1989 itself brought uncomfortable relations with existing major trading partners, especially the USA.

In such new environments in FDI, both the number and value of Korean FDI projects increased dramatically after the mid-1980s. In particular, the number and value of Korean FDI projects in the EU have shown an increasing trend since 1986. In comparison with the pattern of Korean FDI in other regions, the pattern in the EU member nations was quite different. Korean FDI in the EU has been dominated by consumer electronics firms.

The dramatic increase in the number and value of Korean FDI in the EU was made because, first, Korean firms changed their strategy from export-oriented methods to local-production-oriented methods. In fact, Korean firms changed their strategy because the EU started to restrict their exports by using various tariff and non-tariff barriers.

3

THE POLICY AND PERSPECTIVE OF THE KOREAN GOVERNMENT TOWARDS OUTWARD FDI

The strategy of investors can be influenced by both the investment measures of host countries and the overseas investment policies of home countries. Depending on the diverse policies of host countries, foreign investors have been encouraged or discouraged. For example, host governments can encourage foreign investments by the simplification of the procedures and the acceleration of decisions concerning investment approvals (Wells and Louis 1983: 144). They can also discourage foreign investments by complication of the procedures and the delay of decisions concerning investment approvals. The overseas investment policies of home countries have been almost as diverse as those of host governments (Wells and Louis 1983: 144). Depending on the diverse policies of home countries, their firms should set up their foreign investment strategy. For example, Mexico and Hong Kong do not require approval for their firms to invest abroad, while some other countries (Korea and Taiwan, for instance) require potential investors to obtain permits. Some countries, such as India, have fairly explicit rules. The Indian government requires its firms to enter joint ventures when they go abroad, and in the past required them to limit contribution to Indian machinery and to export no foreign exchange (ibid.). Most home governments, as is the case in Brazil, for example, appear to put primary importance on any exports that a project might generate from the home country or any access to raw materials needed at home (ibid.).

From 1968 until February 1989 the Korean government required Korean firms to seek approval for each project. Since 1968, when the first Korean overseas investment was made, the Korean government's policy towards FDI has been characterised by frequent relaxation and liberalisation of the Foreign Exchange Control Regulations (FECR) on which Korean policy towards FDI has been based since the early 1960s in order to cope with the growing need for overseas activities. Particularly after 1986, the government's investment policy showed a more permissive attitude. Internal and external changes in the surrounding economic environment have been responsible for that trend. The Korean government has changed or created several policy measures to cope with the growing need for overseas activities of firms.

As an example, on 30 July 1992, the Korean government announced the changes in all 550 articles of FECR, which were effective from 1 September 1992 (*NaeWae KyungJae ShinMun* (newspaper) 30 July 1992). Part 15, which is on foreign direct investments, was also changed. As recently as 1986, however, the government had adopted a quite restrictive position in its policy on outward direct investment by Korean firms because of the chronic deficit in the balance of payments position and the heavy burden of foreign debt.[1]

The purpose of this chapter is to understand how the policies of the Korean government have played a role in investment decisions by consumer electronics firms in local production in the EU. Thus, a history of the historical development of part of the FECR, and the present regulatory systems, are reviewed. This review has three sections. The first section is on the historical development of Korea's Foreign Exchange Control Regulations (Part 15). The second section is on the present regulatory systems. The third section is a conclusion.

3.1 Historical development of the foreign exchange control regulations (Part 15)

The Foreign Exchange Control Act (FECA),[2] which was promulgated on 31 December 1961, and its Enforcement Decree[3] were the main laws preventing outward investment by Korean firms. FECA strictly and comprehensively controlled foreign exchange transactions and commercial transactions of domestic and foreign firms related to the import and export of foreign currency. The Korean government imposed extensive controls on foreign exchange transactions and commercial transactions in order to avoid deficits on its balance of payments, and thus to maintain the value of the won, and make sure there was sufficient foreign exchange to pay for imports (FECA, Law No. 933, Article 1).

In 1964 the Korean government enacted the FECR[4] based on the FECA (FECA, Law No. 933, Article 23) and its Enforcement Decree (Cabinet Decree No. 382, Articles 33, 35 and 37). However, the FECA did not include Part 15[5] until December 1968, when Korea's first foreign investment was made for a timber project in Indonesia by a Korean firm (P.T. Development Co. Ltd). Before 1968, the Korean government did not allow any individuals or legal entities to make any form of foreign investment, and none asked for an approval before then. Part 15 of the FECR has 17 articles which provide for definitions, the types of FDI, application requirements, procedure for approval and reporting requirements (Shin and Oh 1990: 32).

Between 1968 and 1993, however, government policy shifted from total opposition to outward investment to active encouragement of it. This was done in five stages: the incipient stage, the stage of developing the FDI system, the stage of strengthening the supervision of FDI, the stage of the

Stage I: Incipient stage (1968–April 1975)
No FDI by Korean firms until 1968.
New regulations on outward FDI appear in foreign Exchange Control Law.

Stage II: Stage of developing FDI system (April 1975–October 1978)
Stipulation of 'Guidelines on FDI permit and follow-up procedures'.
a) Responsibility for FDI permits for and follow-up procedures assigned to the Bank of Korea.
b) FDI firms should file reports on overseas activities.
c) Governmental representatives, such as consulates, in the respective foreign countries are assigned to supervise the FDI operations.

Stage III: Stage of the strengthening the supervision of FDI (October 1978–October 1980)
Strengthening supervision of FDI permit and follow-up procedures.
a) Qualification for FDI has been tightened.
b) Prior approval system has been adopted on the operation budget of FDI business.
c) Classification of FDI permit is divided into three categories, i.e., probibition, suppression, and encouragement.

Stage IV: Stage of the vitalisation of FDI (October 1980–1986)
Need for FDI vitalisation has been increased.
a) Protectionist measure of importing countries.
b) Expansion of resource nationalism

Strategy of FDI vitalisation
a) Ease of FDI qualification.
b) No limit on host countries.
c) Abolish system of prior approval on the operational budget of FDI business. Instead of prior approval, a FDI screening committee is installed.
d) The category of FDI suppression business lines are deleted.
e) Regulations on withdrawal of principal within ten years are deleted.

Stage V: Stage of the liberalisation of FDI (1987–)
a) In January 1987, the government adopted a simple approval system based on meeting the requirements which it has set out with respect to foreign investment involving an amount of less than US$ 200,000, and technology service projects.
b) In April 1987, the government raised the minimum amount to be subjected to the Foreign Investment Deliberation Committee for corporate investors from US$ 1 million to US$ 3 million. In December 1987, the government raised again the minimum amount from US$ 3 million to US$ 5 million. An individual investor's minimum amount in this regard was also updated from US$ 500,000 to US$ 1 million
c) The government eased the regulations for real estate purchases by Korean firms abroad.
d) In November 1988, the minimum amount of investment previously subject to automatic approval was raised from US$ 1 million to US$ 2 million.
e) In February 1989, the government abolished a corporate point system.
f) In September 1992, the minimum amount of investment previously subject to automatic approval was raised from US$ 2 million to US$ 5 million.
g) In September 1992, the government raised the minimum amount to be subjected to the Foreign Investment Deliberation Committee for investors from US$ 5 million to US$ 10 million.

Figure 3.1 Historical changes in FDI regulations

Source: Table 1 in *Foreign Direct Investment From Developing Countries: The Case of Korean Firms* (Euch and Min 1986: 153), *Uri NaRaeu HaeWaeGikJupTuJawha KwalRunJaeDoeu GaeSun-BangHang {Korean FDI and Improvement of its FDI-related Regulations}* (Han 1996).

vitalisation of FDI and the stage of the liberalisation of FDI. As illustrated in Figure 3.1, the five stages are classified according to the overall policies adopted by the Korean government.[6] At the incipient stage, from 1968 to April 1975, the Korean government severely restricted Korean investment abroad because of large deficits in the balance of payments and the heavy burden of foreign debt. The only exceptions were projects like forestry and fisheries that were deemed absolutely necessary to ensure secure supplies of raw materials needed for national economic development.[7] The Korean government previously attached the main emphasis of foreign investment policy to the formation of international joint venture enterprises which would secure stable supplies of the natural resources which Korea needed. Especially favoured, therefore, were those investments that involved the development and subsequent importation of such natural resources as coal, oil, gas, and forestry and fishery products.

At the stage of the developing FDI system, from April 1975 to 1979, there was the expansion of the internationalisation of Korean enterprises. During this period, Korea's general trading companies very actively established many small-scale corporate sales affiliates abroad (Shin and Oh 1990: 33). In April 1975, the Ministry of Finance (MOF) amended Part 15 of the FECR for the first time in order to set up a governmental system of streamlining FDI. The Korean government set forth a more detailed approval system which was a positive step to encourage Korean firms to invest abroad by stipulating 'guidelines on FDI permission and follow-up procedures'. According to the new Part 15, as one of the important changes, the approval authority for foreign investment was formally delegated to the Governor of the Bank of Korea.

Under the delegated authority, the Foreign Investment Deliberation Committee was formed in the Bank of Korea in 1976 in order systematically to oversee Korea's foreign investment activities. Those steps included significant tax incentives, low-cost finance, double taxation agreements and other supportive measures. In fact, Korea's general trading companies established many corporate sales affiliates in foreign countries, even though most of them were newly formed and on a small scale.

At the next stage, from October 1978 to October 1980, the government strengthened supervision on FDI permit and follow-up procedures. The government tightened the qualification for FDI and adopted the prior approval system on the operation budget of FDI business. The government also made the classification of FDI permits by dividing them into three categories, i.e., prohibition, suppression and encouragement. In more detail, the Korean government clearly outlined the policies for overseas investments by its firms. It specified the following four areas in which 'overseas investment shall be promoted and supported': (1) 'investment for development and import of raw materials essentially required at home'; (2) 'investment for overcoming any bottleneck in exports';[8] (3) 'fisheries investment to secure fishing grounds' and (4) investment in industries in which 'competitiveness in the world market has been weakened under the nation's industrial structure'.[9] The government strictly prohibited investments 'which would have serious adverse effects on the Korean economy'. The Korean government also prohibited the investment in the form of 'emigration funds'.[10]

At the stage of the vitalisation of FDI, from October 1980 to 1986, the government took several measures such as relaxation of the qualification for FDI and abolition of the system of prior approval on the operational budget of FDI business, because the need for FDI vitalisation had been increased. From the beginning of the 1980s, many countries, especially developed countries, started to take protective measures and to place greater importance on the protection of their national resources. As an example, in the early 1980s, the French government made great efforts to produce the European Union's New Commercial Policy Instrument which was designed to facilitate action against illicit commercial practices by non-EU member states.

At the stage of liberalisation of FDI, from 1987, the government attempted to streamline the foreign investment procedure even further. The MOF adopted a simple approval system in January 1987 based on meeting the requirements which it had set out with respect to foreign investment involving amounts less than US$ 200,000 (MOF Notice No. 86–23 1986) and technology service projects[11] regardless of amount. Encouraged by the surplus in the balance of payments, achieved by the Korean government for the first time in 1986, the minimum amount subjected to the Foreign Investment Deliberation Committee for corporate investors was raised once again from US$ 1 million to US$ 3 million in April 1987. In December 1987, the minimum amount was further raised from US$ 3 million to US$ 5 million. According to MOF Notice No. 87–20, which was announced on 26 December 1987, an individual investor's minimum amount in this regard was also updated from US$ 500,000 to US$ 1 million. The Korean government was determined to enhance foreign business activities as a way of spending the foreign currency accumulated in Korea. Thus, Korean businesses incorporated abroad were allowed to retain abroad up to US$ 500,000 of their net profits or 20 per cent of the initial investments without the permission of the Bank of Korea.[12] In addition, the deliberation period of the Foreign Investment Deliberation Committee for approval of foreign investment projects was shortened from 20 days to 10 days.[13] Another component of the new measures involved an easing of the regulations for real estate purchases by Korean firms abroad. Korean construction firms incorporated in foreign countries were allowed to engage in real estate development projects, such as the construction of hotels, apartment complexes and shopping facilities for resale or rent.[14] The year 1987 also saw the emergence of an automatic approval system by notification for foreign investment if the investment was less than or equal to US$ 1 million (MOF Notice No. 87–20 1987). Korea was suffering from high inflation as well as labour unrest for most of 1987, and foreign investment was thought to be a remedy, not only for inflation but also to make Korean products more price-competitive abroad. Small- and medium-sized Korean firms began investing in Southeast Asia, attempting to explore low-cost labour markets and to penetrate the export market efficiently. As a result of a flood of small- and medium-sized Korean businesses investing abroad, the MOF relaxed the points system whereby small- and medium-sized Korean firms had to score 45 points (from the previous 55) in order to receive permission to invest from the Bank of Korea (Bank of Korea 1989: 5). Individuals' foreign investments exceeding US$ 1 million were also permitted with prior approval from the Bank of Korea (without being required to be passed by the Foreign Investment Deliberation Committee) from March 1988 (MOF Notice No. 88–8 1988). These new regulations relaxed the prior restrictions on individuals' investment into foreign businesses, which had been limited to less than US$ 1 million and subject to the approval of the Bank of Korea.

In July 1988, Korean investors were no longer required to source investment funds from their own assets; they could also borrow in foreign money markets (Bank of Korea 1989: 6). Royalty collections resulting from Korean technology transfer to foreign countries were no longer considered foreign investment (ibid.). The list of investments encouraged by the Korean government was also deleted from the MOF policy guidelines and Koreans could now invest in areas of their choice.[15] The MOF delegated its authority to approve foreign investments to the President of the Korea Export-Import Bank, for cases in which this Bank loaned foreign investment funds.[16]

In November 1988, according to FECR Article 15–5(1)(i), the minimum amount of investment previously subject to automatic approval was raised from US$ 1 million to US$ 2 million. In February 1989, the MOF loosened the foreign investment regulations further by deleting a corporate point system that was previously used in order to estimate the eligibility of Korean firms (ibid.). According to FECR 15–3(2)(ii), direct equity investment (non-portfolio investment) was further relaxed. Thus, Korean firms could invest in foreign securities if they acquired a minimum of 20 per cent of the voting stocks of foreign firms or if they appointed at least one full-time director to serve on the board of the foreign firm. Previously, a shareholding ratio of 50 per cent or higher, or a veto right in the board of directors, had been required in order to get the Bank of Korea's approval (ibid.).

As of February 1989, minimum interest rates for funds loaned abroad were abolished (ibid.). It was at this time that the Korean government also officially commenced permitting investment in countries having no diplomatic relations with Korea. With the exception of investments in socialist countries, a significant change in Part 15 of the FECR is the adoption of the automatic approval system whereby Korean investors with investments of less than US$ 2 million can now simply report to the Bank of Korea on foreign investment (FECR Article 15–5 (I) (ii)).

On 30 July 1992 the Korean government announced that all 550 articles of FECR, including Part 15 which is on foreign direct investments, would be abolished from 1 September 1992 (*NaeWae KyungJae ShinMun* (newspaper) 30 July 1992). According to the changed regulations, the maximum investment amount which is eligible for the simple notification system was raised from US$ 2 million to US$ 5 million from the beginning of September 1992.

The government also changed the maximum overseas investment amount which required approval from the Foreign Investment Deliberation Committee. Before September 1992, all projects had to be examined by the committee if the projects were more than US$ 5 million or made in countries without diplomatic relations with Korea. According to the new regulation, projects are required to be examined by the committee if they are more than US$ 10 million or made in a prohibited area of industry.

3.2 Regulatory systems in the 1990s

Since the change announced on 30 July 1992, there have been four more changes of policy on outward direct investment.[17] The current overseas investment policies of the Korean government offer a set of positive incentives for overseas direct investment as it becomes increasingly a necessity: (1) expanding the automatic approval system; (2) reinforcing financial assistance including loans and guarantees, and tax incentives; (3) protection from investment risks and (4) an information service.

First, expanding the automatic approval system was a positive incentive for overseas direct investment. Before February 1989, prospective Korean investors had to apply for and obtain permission from the governmental Foreign Investment Deliberation Committee at the Bank of Korea when they wanted to invest overseas. The approval system in the 1990s is different. The Korean government plans eventually to abolish the restrictions in overseas investment for full liberalisation. The government raised the maximum investment amount that is eligible for the simple notification system from US$ 10 million to US$ 50 million from the beginning of October 1995. Thus, investments not exceeding US$ 50 million are possible simply by notifying the permission office concerned, without prior approval from the government. However, investment exceeding US$ 50 million or any investments in a restricted industry still require approval from the permission office concerned. Previously, approval was required if investments exceeded US$ 10 million or if they were made in countries without diplomatic relations with Korea.

The second positive incentive is the reinforcement of financial assistance and tax incentives. First, financial assistance to Korean investors was provided in the form of medium- and long-term loans and guarantees by the Korea Export-Import Bank as well as in the form of standby credit from other Korean banks, as Tables 3.1 and 3.2 illustrate. Until the 1990s, the government

Table 3.1 The outward direct investment support system of the Export-Import Bank of Korea (as of July 1997)

	Overseas investment credit
Eligible borrower	Korean investors
Credit coverage	Up to 90% of the necessary funds to implement overseas investment
Interest rate	Fixed: Commercial Interest Reference Rate (CIRR) ~ CIRR + 1% per annum (maximum). Floating: LIBOR + (0.5 ~ 1.25%) per annum.
Maturity	Up to 10 years, including grace period depending on business plan. Grace period: up to 3 years. Commitment fee: Large-size firms: 0.5% per annum. Small- and medium-size firms: n.a.
Repayment	Annual or semi-annual instalments.

Source: Export-Import Bank of Korea.

Table 3.2 Overseas investment loan of Economic Development Co-operation Fund (EDCF) (as of July 1997)

Eligible borrower	*Korean investors*
Loan coverage	Up to 80% of the necessary funds
Interest rate	1–5% per annum
Maturity	Up to 30 years, including grace period
Grace period	Up to 10 years
Repayment	Semi-annual instalments

Source: Export-Import Bank of Korea.

had been increasing the annual ceiling of overseas investment credits financed by the Korea Export-Import Bank[18] as well as granting more favourable terms and conditions, especially for small- and medium-sized firms. However, with the Gulf War and the deterioration of the balance of payments, the bank temporarily reversed the trend in 1991.

Overseas investment credit is available to almost all prospective projects except the purchase of real estate[19] or unproductive industry.[20] The credit covers up to 90 per cent of the investment value for a maximum repayment period of 10 years with a fixed or floating interest rate, as Table 3.1 shows. It also extended the facilities and services for overseas investment credits to previously ineligible industries and firms.

Second, the Korean government also provides some tax incentive programmes to firms which reserved funds for possible investment losses, and ensures there is no double taxation on profits and incomes earned abroad. In more detail, tax incentives are offered to Korean investors in overseas operations through such measures as deductions of taxes paid to foreign governments by Korean investors, reduction of tax rates on incomes and corporate earnings from overseas direct investment, and exemption from double taxation.

The third positive incentive was protection from investment risks by extending substantial support for overseas investment insurance. As illustrated in Table 3.3, the Korea Export Insurance Corporation (KEIC)[21] has offered overseas investment insurance on behalf of the government to protect Korean investors from possible uncertainty and instability associated with political turbulence such as war, expropriation or inconvertibility.

Table 3.3 Overseas investment insurance (as of July 1997)

Eligible insured	*Korean investors*
Percentage of cover	The maximum coverage is 90% of the invested amount plus profits earned (maximum 95% for small- and medium-size firms).
Period of cover	Up to 10 years.
Indemnity	90% of the invested amount plus profits less the amount repatriated prior to the occurrence of loss (maximum 95% for small and medium-size firms).
Premium rate	0.18%–1.44% of insured amount per annum. according to country risk and risk covered.

Source: Korea Export Insurance Corporation.

The KEIC will improve overseas investment insurance as part of its ongoing plan to enhance the programme. However, the KEIC only provides overseas investment insurance to protect Korean firms from losses resulting from equity investment, long-term capital outflow, public or private bonds acquisition or real estate investment. This insurance covers up to 90 per cent of the invested amount plus profits for up to 10 years with the premium rate of 0.18–1.44 per cent. To help minimise the risk of losing invested capital in socialist nations,[22] the Economic Planning Board also introduced a mandatory approval system on 1 April 1989.

Fourth, the government tries to make investment protection agreements with more foreign countries. The government plans to create a favourable and secure atmosphere for Korean firms' overseas investment by increasing the number of countries entering into agreements for the promotion and protection of investment and/or anti-double taxation, etc. As of May 1995, 33 countries entered into the former and 45 countries entered into the latter.

The fifth positive measure is gradual liberalisation of controls over overseas indirect investment. Furthermore, the government plans to make overseas indirect investment free, step by step, in order to prepare for the opening of the domestic capital markets and heighten the ability of Korean firms to manage their international portfolios. Currently, only institutional investors, such as securities firms, investment trust companies and insurance companies, are allowed directly to invest in foreign securities. By restricting individual investment, it is hoped that these financial institutions will accumulate experience in international portfolio management and augment the efficiency of asset management by diversifying their investment portfolios. The government plans to allow corporations and individuals to invest directly in foreign securities in the late 1990s or early next century, during the Seventh Economic Development Plan period.

The sixth positive measure is providing sufficient investment information. The Korean government assisted the Export-Import Bank in setting up an Overseas Investment Research Institute on 15 November 1988 to provide overall services to Korean firms regarding overseas investment information and consultation to potential investors. This institute comprises the Economic Research Department, Country Studies Department and the Overseas Investment Information Centre.

Overseas investment information has been gathered from such diverse sources as government agencies, branch offices of Korean banks, and trading companies. The basic weakness in the information-gathering activities is that there is no governmental or quasi-public agency especially designed for the collection, analysis and dissemination of overseas investment information.

3.3 Conclusion

The Korean government is, in the late 1990s, reassessing its strategies concerning investment for future economic development against a trend of growing protectionism and regionalism. The Korean government has eliminated various restrictions on foreign exchange transactions relating to foreign investment since the mid-1980s. This measure reflects the demand of firms that the government support them by relaxing its regulations on foreign exchange transactions relating to foreign investment, showing its confidence in the liberalisation of the Korean economy and its intention to lay the foundation for full deregulation of capital transactions. In fact, the FECR has been the regulation that has been changed most frequently in Korea since 1986.

Proper management of the FECR or new legislation specifically designed for foreign investment activities may be able to contribute to further rapid economic development in Korea. It is also very important in Korea's foreign investment activities that the government promotes FDI through intergovernmental investment protection agreements as well as various types of financing and governmental incentives.

4

THE RELATIONSHIP BETWEEN GOVERNMENT AND BUSINESS IN KOREA

Since July 1963, when Korea established diplomatic relations with the EU, political relations between the two have been strengthened to the extent that they have exchanged delegations.[1] Economic relations between Korea and the EU have also improved over recent decades. Nevertheless, it is only very recently that bilateral economic co-operation has been added to the agenda of trade diplomacy. However, it is still true that for both Korea and the EU, economic relations with the US and Japan are much more important than those with each other. Since the 1992 Single Market Programme was announced in 1985, both Korean FDI and exports to the EU have increased dramatically, and in particular, as discussed in Chapter 2, Korean FDI in the EU member states has increased since 1986. There were many internal and external factors that brought about this increase. One internal factor was the Korean government's policy supporting Korean firms investing in the EU. In Chapter 3 it was briefly explained how the Korean government controlled firms investing in the EU.

However, there was no explanation of why and how Korean firms were influenced by Korean government policy when they were considering FDI in the EU, or why and how the Korean government was influenced by Korean firms. In other words, the relationship between the government and business in the decision-making process of Korean FDI in the EU is yet to be explained in detail.

It is generally believed that the relationship between the government and business in Korea has a superior-subordinate nature (Cho and Hellmann 1993: 10; Steers, Shin and Ungson 1989: 47). However, according to the results of the questionnaire and of interviews with senior managers of Korean firms in the EU, it is clear that it was Korean firms who took the initiative, rather than the Korean government, in the decision-making on the issue of FDI in the EU.[2] The government would seem simply to have acceded to requests from businesses considering and making FDI in the EU. The question, then, is 'Why did the Korean government act in this exceptional way in its relations with firms?' In fact, this is part of a wider story, because most governments have similar relations with firms in their countries, although there are some differences in the degree to which firms take the initiative.

In this chapter, therefore, first, the general relationship between the government and business in Korea is analysed in comparison with those in Japan and Taiwan. Japanese firms are often regarded as the model for Korean firms, both directly and indirectly. For instance, the Korean government actively pushed successful Korean firms to hasten their expansion, after observing how the Japanese government in the 1960s had similarly prepared for the inevitable opening of its markets as a result of foreign pressures. At that time the Japanese government tried to increase the size of Japanese companies in order to enable them to produce competitive goods on a large scale and thus keep out imports through market forces. Taiwan is included because of similarities between the Korean and Taiwanese economies.

Second, I will analyse why the Korean government acted in an exceptional way in its relations with those firms considering and making FDI in the EU.

4.1 The general relationship between government and business in Korea in comparison with those in Japan and Taiwan

4.1.1 The general relationship between government and business in Korea

The general relationship between government and business in Korea has remained unchanged since 1961. In short, as Figure 4.1(A) shows, since 1961 the relationship has had a decidedly superior–subordinate nature compared to Japan (Cho and Hellmann 1993: 10; Steers, Shin and Ungson 1989: 47). In Korea, government has set the policies and, from time to time, has ordered business to do certain things and not do others, and firms have typically done as they were told (Jung 1989: 12; Smith 1979: 17). In more detail, the relationship between government and business in Korea for the period from 1953 to 1995 has gone through four distinct phases; from 1953 to 1960, from 1961 to 1979, from 1980 to 1992, and the period from 1993 to 1995.

(i) The relationship between government and business between 1953 and 1960

Before 1961 relations between the two were very different. During the period from 1953, when the Korean War ended, to 1960, when the government of President Rhee Syng-Man was brought down by the student revolution, government–business relations were quite different from the 1960s, 1970s and 1980s. Before 1961, the government did not display strong-handed or heavy interventionist characteristics in relation to business for the following reasons. First, as Jung (1989: 14) points out, President Rhee himself was a

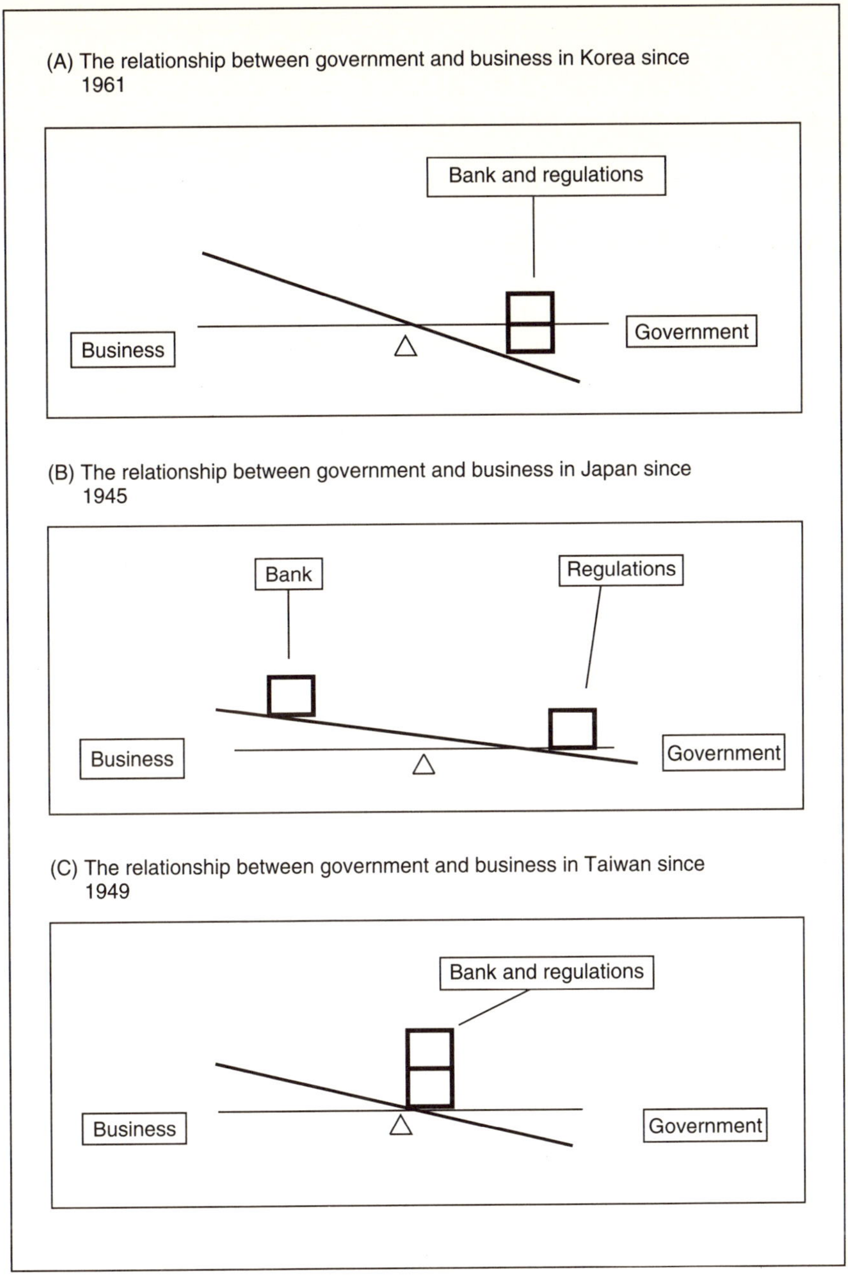

Figure 4.1 The relationship between the government and business in Korea (since 1961), Japan (since 1945) and Taiwan (since 1949)

believer in free market economy. He was 'an independence leader who devoted his attention to politics and national integration while ignoring economics' (Jones and Sakong 1980: 290). Although the President did not pay as much attention to the economy as to politics and national integration, he none the less respected 'the principles of a competitive, private market economy' (Jung 1989: 14). Second, in the early period of the Republic, the US military administration exercised a huge influence on the economic policy-making of the Republic of Korea. Third, the Korean government did not then have suitable institutions and know-how to control the economy. Last, the people might have reacted adversely to any form of control because of the painful memory of the controlled economy during 36 years of the Japanese occupation.

The development strategy of the Korean government in the 1950s, which was neither well articulated nor well planned, was based upon import substitution (Dicken 1992: 177), overvalued exchange rates and heavy reliance on foreign assistance (Mason *et al.* 1980: 249). However, it cannot be denied that, in some respects, the developments of the 1950s laid the foundation for later industrial growth.

(ii) The relationship between government and business between 1961 and 1979

A new pattern of government–business relationship started from the birth of the Third Republic of President Park Chung-Hee. Hasan explained the relationship between government and business in the 1960s and 1970s as follows:

> The apparent paradox is that the Korean economy depends in large measure on private enterprise operating under highly centralized government guidance. In Korea the government's role is considerably more direct than that of merely setting the broad rules of the game and influencing the economy indirectly through market forces. In fact, the government seems to be a participant and often the determining influence in nearly all business decisions.
>
> (Hasan 1976: 29)

The relationship between government and business in the 1960s and 1970s was very close[3] and contributed greatly to economic growth. This pattern of government–business relationships under President Park Chung-Hee was set in the first few months of his regime. Instead of punishing profiteers, as under the Rhee government, President Park formed an alliance with business that laid the basis for subsequent economic development. In other words, President Park put a high priority on economic growth that was second only to the national interest and his personal political survival (Jones and Sakong 1980:

290). One of the earliest and most important measures of the new government was the implementation of the Special Law for Dealing with Illicit Wealth Accumulation.[4] Because of the Special Law, many leading Korean businessmen were arrested and threatened with confiscation of their assets. Soon after the law was passed, 10 business leaders were summoned to a meeting with Park Chung-Hee, then still Vice-Chairman of the Revolutionary Council. The meeting ended with the following three agreements: (1) 'the government would exempt most businessmen from criminal prosecution'; (2) 'with the notable exception of commercial bank shares, existing assets would not be confiscated' and (3) 'businessmen would instead pay off their assessed obligations by establishing new basic industrial firms and donating a share to the government'[5] (Mason *et al.* 1980: 262).

One of the representative events showing close and co-operative relations between the government and business in the 1970s was the agreement of the Korean government to the reduction of corporate tax rates and rates of interest on bank loans on 3 August 1972[6] (Ogle 1990: 43). At the request of the chairmen of the *chaebol*s, President Park reduced the rates of interest on the debts owed to the banks in order to prevent an economic collapse which would have been disastrous for the President himself. At that time, President Park was in a very difficult situation; popular protests and demonstrations against him were common. People protested strongly against him because, in November 1971, he had declared a national emergency and arrogated all government power to his own hands. To avoid making the situation worse, it was vital to prevent economic disruption. To do this, as mentioned earlier, Park lowered the real interest rates on bank loans at the request of companies because they had always operated under a heavy load of debt. Owing to an interest rate reform in 1965 that was designed to mobilise household savings through formal monetary financial institutions, essentially by maintaining significantly positive real interest rates on time deposits, there were dramatic increases in bank savings as well as in the money–GNP ratio itself, as Table 4.1 shows. This implies an increased supply of loanable funds through the formal institutions.

In addition, the Korean government's implementation of the Special Law for Dealing with Illicit Wealth Accumulation left a strong imprint on subsequent government–business relations in Korea and also on the social image of *chaebol*s (Jung 1989: 15). As Jung (ibid.) points out, first of all, people gained a very negative image of *chaebol*s because the businessmen concerned had to admit publicly that they had evaded taxes and profited from unfair and illegal deals, regardless of how much they had been used by the government during this politically unstable period.

Second, the superior position of the political elite relative to businessmen was firmly established in Korea in the course of this settlement. Chairmen of *chaebol*s confessed to having committed illegal acts but the government forgave them and, furthermore, invited them to take part in a subsequent

Table 4.1 Savings, investment and consumption, as percentage of GNP (1962–84)

Year	*1*	*2*	*3 (= 1+2)*	*4*	*5*	*6*	*7(=4 +5+6)*	*8*	*9(= 1-4)*
1962	13.39	−0.90	13.04	4.63	7.91	−0.97	11.58	1.86	−1.47
1963	13.94	4.43	18.38	4.41	7.76	3.45	15.63	4.05	−2.75
1964	11.56	2.97	14.53	4.61	7.00	3.61	15.22	0.84	0.69
1965	14.60	0.10	14.70	5.83	8.07	0.18	14.09	−0.16-a	−0.61
1966	19.96	1.66	21.62	5.68	7.72	4.15	17.56	2.72	−4.06
1967	21.25	0.66	21.91	6.83	8.16	1.38	16.36	4.18	−5.55
1968	25.54	1.20	26.74	8.17	8.19	3.06	19.43	7.73	−7.31
1969	26.74	3.29	30.02	7.36	7.97	7.45	22.78	7.73	−7.25
1970	25.60	2.39	27.98	7.62	7.67	3.51	18.80	7.59	−9.18
1971	21.52	3.61	25.13	5.57	7.53	3.17	16.27	8.73	−8.86
1972	20.00	2.22	22.22	3.64	8.72	5.70	18.06	3.48	−4.16
1973	23.38	2.30	25.68	3.99	11.17	8.98	24.14	2.29	−1.54
1974	25.31	6.34	31.65	2.25	11.36	7.04	20.65	10.93	−11.00
1975	25.50	4.53	30.02	3.76	9.81	6.61	20.18	9.05	−9.84
1976	24.09	1.54	25.62	6.05	10.21	8.81	25.07	1.09	−0.55
1977	26.66	1.08	27.75	5.11	10.68	12.30	28.09	−0.03	0.34
1978	30.81	0.38	31.19	6.17	9.94	13.28	29.40	2.17	−1.78
1979	32.77	2.88	35.65	6.74	9.75	12.29	28.78	6.43	−6.87
1980	31.91	−0.66	31.26	5.67	10.29	6.63	22.59	8.67	−8.67
1981	28.85	0.29	29.15	6.12	9.34	6.96	22.42	6.91	−6.73
1982	30.27	−3.27	27.00	6.18	9.69	7.23	23.09	3.77	−3.90
1983	30.84	−4.07	27.77	7.46	10.28	7.89	25.63	2.09	−2.14
1984	30.95	−0.97	29.98	7.64	10.43	9.93	28.00	1.68	−1.98

Source: *Asia's Next Giant: South Korea and Late Industrialization* (Amsden 1989: 75, Table 3.7).

Notes: Negative value equals surplus. *1*: Gross fixed capital formation. *2*: Increase in stocks. *3*: Total investment. *4*: General government savings. *5*: Public and private corporations' savings. *6*: Household and private non-profit institutions' savings. *7*: Total savings. *8*: Deficit of the nation on current account. *9*: Difference between investment (*1*) and savings (*4*).

economic development process. Thus, everybody knew who was in the driver's seat in the relationship between the government and business.

Third, after this event, businessmen who were accused of illegal profiteering founded a lobbying organisation which later became the powerful Federation of Korean Industries. Through this organisation, businessmen increased their voice in relations with the government, especially in the late 1980s and 1990s.

Fourth, according to the second agreement between the government and business in the process of settlement of the Special Law for Dealing with Illicit Wealth Accumulation, five major nation-wide commercial banks that had been controlled previously by private businessmen were put under the control of the government. This laid a foundation for the future control of bank credit by the government.

Of these four results from the implementation of the Special Law, it was the government's ownership of banks that particularly affected the future relationship between government and business in Korea. Owing to this, first of all, the Korean government could virtually completely control private business activities, mainly through such financial and monetary means as preferential loans and interest rates, and through control of access to foreign loans. In Korea, in fact, the government used its power through financial monetary and other means – such as licensing authorisations and inclusion of certain companies in its recurring five-year plans – to select and then guide those chosen for success (Steers, Shin and Ungson 1989: 43). All the leading observers of Korean economic development agree that, from 1961, the most potent instruments for implementing government-led economic policy have been state control of bank credit and access to foreign borrowers (Johnson 1987: 147; Luedde-Neurath 1988: 75; Mason *et al.* 1980: 267; Sakong 1993: 72).

It was possible for the government to control bank credit in Korea because the government owned all five major nation-wide commercial banks,[7] six special banks and two of the three development institutions[8] that undertook corporate financing until the early 1980s (Cole and Park 1979: 56). In addition, the government's bank ownership meant that it could even dictate the objectives of bank operations, lower-level personnel policy and budgets. Thus, the government could also appoint senior bank managers, so that a manager's most important task was to conform to the policy guidelines established by the government. In fact, managers of banks carefully executed government directives and prudently avoided visible failures such as large loan defaults.[9] The government even set credit ceilings for individual banks, controlled their operating funds and, as previously mentioned, their interest rates (Hong 1979: 40; ILT 1980: 22, 23; ILT 1983: 25; Sano 1977: 52).

The government's bank ownership was also significant for firms because credit was particularly critical in Korea, where the debt-equity ratios in manufacturing had been in the range of three or four to one in the first half of the 1970s.[10] In other words, *chaebol*s strongly depended on bank loans for the expansion of their business, as Table 4.2 illustrates.

During this period, the government, as in most LDCs, employed tight controls on bank interest rates and directed loans to support specific sectors and industries. In fact, the Park government was 'in favour of borrowing from abroad and keeping domestic interest rates low in order to increase investment' (Amsden 1989: 73). The result was excess demand, creating the need for credit rationing. This introduced a command feature into what would have been, if interest rates had governed access to credit, an incentive system. During the Park government, the rationing of credit was highly discretionary and frequently also politically motivated, just as it was under the Rhee government. Under Park there had been a serious attempt at specifying desirable end uses, differential rates to be allowed and target industry allocations. In particular, the Korean government allocated more than half of all bank credit

Table 4.2 The ratio of loans from banks to total loans of *chaebols* (1987–9) (percentage)

	1987	*1988*	*1989*
5 largest *chaebols* (*1*)	73–76	70–69	75–51
10 largest *chaebols*	73–60	71–74	74–13
20 largest *chaebols*	75–56	74–31	76–62
30 largest *chaebols*	75–35	74–27	76–47
Sample firms (*2*)	56–00	58–26	59–93

Source: *The Economic Analysis on Korean Chaebols* (Jung and Yang 1992: 105, Table 318)

Notes

1: Samsung, Hyundai, Daewoo, Lucky-GoldStar and Hanjin.

2: Sample firms that were used in the book *The Analysis of Business Management* (Bank of Korea).

to directed programmes (Cho and Hellmann 1993: 2). The Korean government gained huge influence over the direction of corporate expansion through this tight control on bank interest rates and by directing loans to support specific sectors and industries. It was possible for the government to discriminate in favour of particular industries and firms because the cost of borrowing at home was far exceeded by that of borrowing abroad.

As illustrated in Table 4.3, the general bank interest rate had usually been well below the curb rate and, in real terms, it had often been negative. In Korea, there have been two markets for credit: banks and an underground money market (curb market) (Koo and Kim 1992: 130). In the 1960s and the early 1970s, there was a huge underground money market in Korea. According to Kim (1987), in 1964 the estimated outstanding assets and liabilities in the curb market were almost double those of commercial bank loans. Cole and Park (1983: 126) reported that the total volume of informal loans increased by 450 per cent between 1964 and the end of 1969. They also reported (ibid.: 127) that, as of 3 August 1972 'about 25 per cent of total informal debts (of all the business enterprises) were "disguised informal loans", in that they were made by the owners, major stockholders, or executives of the borrowing firms'.

As Koo and Kim (1992: 130) explain, not just professional money-lenders, but propertied middle-class people, business owners and top managers of the same industrial firms that were borrowing money from the curb market participated in the underground money market as lenders. According to Ahn and Jung (1995), as of the end of 1993, the curb market was still large, taking 22 per cent of Korean GNP (58,403.6 billion won). The curb rate fell dramatically from 46.4 per cent in 1971 to 21.4 per cent in 1991, as Table 4.4 shows. This was mainly because Korea was in a much more favourable economic

Table 4.3 Cost of foreign capital (1966–83, annual averages)

	1966–70	*1971–5*	*1976–80*	*1981–3*
I Domestic bank lending rate[a] (curb market interest rate)	24.4 (54.2)	17.0 (40.1)	18.0 (41.3)	12.5 (30.6)
II Foreign interest rate[b]	6.4	7.9	11.5	11.1
III Foreign inflation rate (GNP deflator)[c]	4.9	8.4	5.9	4.1
IV Exchange rate depreciation[d]	5.1	7.8	5.5	10.1
V GDP deflator (rate of change): Korea[e]	14.6	18.7	19.7	9.9
Real foreign interest rate (II–III)	1.5	−0.5	5.6	7.0
Interest rate differential between home and foreign markets (I–II–IV)	12.9	1.5	1.0	−8.7
Real private cost of borrowing abroad $(II + IV - V)$	−3.1	−3.0	−2.7	11.3

Source: *Principal Economic Indicators; Monthly Bulletin* (Bank of Korea various issues), as cited by Park (1985); and Amsden (1989: 92, Table 4.5).

Notes: a: Discounts on bills on deposit money banks (three-year moving averages). b: LIBOR (90 days). c: Average of Japan and United States, d: Bank of Korea standard concentration rate (three-year moving averages). e: Three-year moving averages.

environment (with more capital) for meeting demand for capital in 1991 than in the 1960 and early 1970s owing to its rapid economic development. In the 1960s and early 1970s, in fact, Korea was still a capital-poor country, but its requirements for capital were very high due to the rate of capital investment required for rapid economic growth (Moskowitz 1989: 71). Consequently, the curb rate in the 1960s and early 1970s was much higher than in the 1980s and early 1990s.

The Park government had a second way of exerting influence over firms. It could control firms' access to foreign borrowers. This was possible because corporate borrowing from abroad could be undertaken only with government authorisation and a government guarantee.[11]

There were three mechanisms of foreign credit allocation: (1) public loans by the government; (2) private long-term loans and (3) private short-term credit (Mason *et al.* 1980: 268). Since 1974 public loans that the government made from international agencies or other governments have been covered by the Public Loan Inducement and Management Law. This specified circum stances under which the government guaranteed foreign loans. The second category was private long-term loans which had over three years to maturity. These were covered by the Foreign Capital Inducement Law. Loans had to be guaranteed by either the government, the Korea Exchange Bank or the commercial banks. The third category was private short-term credit, which came

Table 4.4 Interest rates on bank loans and inflation rates (1971–91) (percentage)

Year	*General loan*	*Policy loans[1] (lending rates)* *Export*	*Machinery promotion fund*	*National investment fund*	*Curb market rate*	*GNP deflator*
1971	22.0	6.0	n.a	n.a	46.4	12.9
1972	15.5	6.0	n.a	n.a	37.0	16.3
1973	15.5	7.0	10.0	n.a	33.4	12.1
1974	15.5	9.0	12.0	12.0	40.6	30.4
1975	15.5	9.0	12.0	12.0	41.3	24.6
1976	18.0	8.0	13.0	14.0	40.5	21.2
1977	16.0	8.0	13.0	14.0	38.1	16.6
1978	19.0	9.0	15.0	16.0	39.3	22.8
1979	19.0	9.0	15.0	16.0	42.4	19.6
1980	20.0	15.0	20.0	22.0	44.9	24.0
1981	17.0	15.0	11.0	16.5–17.5	35.3	16.4
1982	10.0	10.0	10.0	10.0	32.8	7.1
1983	10.0	10.0	10.0	10.0	25.8	5.0
1984	10.0–11.5	10.0	10.0–11.5	10.0–11.5	24.8	3.9
1985	10.0–11.5	10.0	10.0–11.5	10.0–11.5	24.0	4.2
1986	10.0–11.5	10.0	10.0–11.5	10.0–11.5	23.1	2.8
1987	10.0–11.5	10.0	10.0–11.5	10.0–11.5	22.9	3.5
1988	10.0–13.0	10.0	10.0–11.5	10.0–11.5	22.7	5.9
1989	10.0–12.5	10.0	10.0–11.5	10.0–11.5	23.7	5.2
1990	10.0–12.5	10.0	10.0–11.5	10.0–11.5	20.6	10.6
1991	10.0–12.5	10.5	10.0–11.5	10.0–11.5	21.4	10.9

Source: *Principal Economic Indicators; Monthly Bulletin* (Bank of Korea various issues); *Korean Economic Indicators* (National Statistical Office, Republic of Korea; March 1992). (Sakong 1993: 244, Table A.18).

Note:[1] Policy loan which was usually earmarked for certain sectors or even specific industries, had preferential interest rates in most cases when compared with ordinary loans.

under the Foreign Exchange Management Laws. 'In contrast to the first two forms, transactions under this law were relatively automatic if the terms conformed to international banking practice. This exception did not weaken the government's control over all foreign borrowing for capital investment' (ibid.).

Cho and Hellmann (1993: 1) explained this relationship between government and business with a participatory mechanism which they called a 'government-led internal organization' (GLIO). They insisted that government, banks and large industrial firms may be said to have formed a GLIO. These explanations were made by considering the government's role in the credit market. They argued that, 'in early economic development, such a participatory approach may have helped overcome pervasive market imperfections. But there were also significant dangers/problems of entrenched interests and institutional inertia' (ibid.). In Korea, as in Japan, the relative importance of GLIO gradually diminished as competitive capital markets and large conglomerates (privately led internal organisations or PLIO) expanded with economic growth. The concept of GLIO is quite vague. For example, there is no explanation of how to define early economic development. The 1960s and 1970s marked a period of economic development in Korea. According to Cho and Hellmann (ibid.), in the 1960s and 1970s the Korean government dominated relations of business with foreign conditions.

(iii) The relationship between government and business between 1980 and 1992

During the period 1980–92 there were some fundamental changes in the relations between the government and business in Korea. First, rapidly increasing foreign debt eventually weakened the government's position in its relations with business. Korean foreign debt increased rapidly throughout this period, as illustrated in Table 4.5. In particular, during 1980–2, when foreign debt increased up to 30 per cent per annum, it could have been a direct threat to the Korean economy; thus it was important for the government to reduce debt levels in order to maintain long-term growth. To reduce foreign debt, the Korean government tried to encourage firms to become more profitable through less direct government intervention, acknowledging the weakened position of the government in its relations with business. In fact, 'the strong influx of foreign loans has contributed heavily to the large foreign debt accrued over the 1980s' (Ogle 1990: 36). According to Amsden (1989: 94), the Korean government has used foreign loans for two purposes: 'to finance its long-term investments[12] and to borrow its way out of a balance-of-payment crisis in order to maintain its long-term growth trend'.

Second, in the early to mid-1980s, economic planners had called for a 'changing role' for the government towards private economic activities; in effect, this meant a reduced role. The government accepted this because it acknowledged that many of the economy's problems during the troublesome economic period of 1979 and 1980 were due to its own intervention in previous years (Jung 1989: 24). In fact, one of the catchwords of the Korean government at the beginning of the 1980s was *min-gan-ju-do*, meaning literally 'initiatives by private sectors'.

Government power has been diffused by partial liberalisation of the financial system in the early to mid-1980s and by import liberalisation (Wade 1990: 321). The partial liberalisation of the financial system became inevitable for the Korean government because the country's economic situation was damaged by the second oil shock in 1979 and by the historic poor harvests in 1980.[13] Because of these two events, many small- and medium-sized enterprises went bankrupt in the early 1980s, which resulted in increased unemployment. The government therefore started taking measures to revitalise

Table 4.5 Korea's external debt position (1961–84)

Year	*Total foreign debt (US$ million)*	*Long-term debt as % of total*	*Total debt as % of GNP*	*Debt services as % of GNP*	*Debt services as % of exports (merchandise exports)*	*Debt service as % of current transaction receipts*[a]
63–71						
1963	157	85.99	4.06	0.05	2.30	1.1
1964	177	94.35	5.29	0.15	4.17	2.4
1965	206	98.54	6.81	0.46	8.00	4.8
1966	392	98.21	10.26	0.34	5.20	2.9
1967	645	89.77	13.62	0.72	10.15	5.3
1968	1,199	92.58	20.07	0.77	9.47	5.2
1969	1,800	89.22	24.07	1.20	13.68	7.8
1970	2,245	83.39	25.48	2.84	28.34	18.1
1971	2,922	83.61	30.06	3.28	28.16	19.7
72–78						
1972	3,589	82.17	33.95	3.87	24.40	18.4
1973	4,260	83.54	31.55	4.35	17.87	14.2
1974	5,937	79.13	32.01	3.25	13.33	11.2
1975	8,456	71.51	40.55	3.38	14.01	12.0
1976	10,533	71.09	36.73	3.50	12.85	10.6
1977	12,648	70.63	33.79	3.58	13.33	10.2
1978	14,871	74.08	29.71	4.16	16.38	12.1
79–84						
1979	20,500	67.80	31.75	4.03	17.68	13.3
1980	27,365	61.22	44.68	4.81	17.13	13.1
1981	32,490	63.80	48.34	5.53	17.98	13.8
1982	37,295	61.94	52.65	6.23	21.15	15.5
1983	40,094	70.58	53.23	6.18	20.07	15.0
1984	43,100	73.55	53.16	6.74	20.75	17.3

Source: *Asia's Next Giant: South Korea and Late Industrialization* (Amsden 1989: 95).

Note: a: Receipts from visible and invisible foreign transactions.

the depressed economy in order to reduce the unemployment rate. One of these measures involved the government trying to sell its own 20–30 per cent shares in five major commercial banks to private individuals in order to revitalise the banking business, which would help to prevent small- and medium-sized firms from going bankrupt. In particular, the Korean government privatised four major commercial banks in the early 1980s. Hanil Bank was privatised in July 1981, Korea First Bank and Seoul Trust Bank were privatised in September 1982 and Choheung Bank in May 1983. However, the government failed to sell its stocks in the five commercial banks to individuals. It therefore forced the banks to sell stocks to small- and medium-sized firms in order to privatise. At the same time, the Korean government implemented special encouraging policies for small- and medium-sized firms in 1982 in order to save them from bankruptcy. As explained earlier, financial liberalisation in the early to mid-1980s made it more difficult for the Korean government to use targeted credit as a policy instrument.

Besides this financial liberalisation, reforms were also attempted in another two areas: industrial policy realignment and import liberalisation. On 31 December 1982, the government abolished the Temporary Law on Banking Facilities which had given the director of the government's Bank Supervisory Board full powers of appointment over those recommended to executive positions of commercial banks, as well as full powers to discharge their executives. The abolishment of the law weakened the position of the Korean government to some degree in its relations with firms.

However, even these changes did not fundamentally change the relationship between government and business, which remained unequal (Luedde-Neurath 1988: 74). This is because the government did not give up its control over firms. For example, the government changed Korean Bank Law on 31 December 1982 (the sixth revisions of the law) (Law No. 3608) in such a way as to continue to control the banks to some degree. The newly revised law included an article (Article 17) which limited any individual shareholder (and related persons) to owning no more than 8 per cent of a bank's shares. 'This provision was introduced primarily to prevent large *chaebol*s from controlling the banks' (Sakong 1993: 72). However, as Table 4.6 illustrates, 'after 1980, when the Government sold its holding in the commercial banks, almost all the shares sold are believed to have ended in the hands of *chaebol* despite stringent measures to prevent this' (Michell 1988: 68).

In addition, import liberalisation in the early to mid-1980s also reduced the government's bargaining leverage with the big firms (Wade 1990: 321). Korean firms had done their businesses in Korea under the government's special protection from foreign competitors and tried to follow government instructions as far as possible in order to take advantage of this special protection.

The Korean government resumed import liberalisation in 1981 and announced a gradual import liberalisation schedule again in 1984, which was extended in 1985 to cover the period 1986–9.[14] This programme was introduced again to increase efficiency in domestic industries (Sakong 1993: 83, 172). In fact, it became indispensable for the Korean government because of its position in the world economy. 'Korea's voluntary import liberalization raised an important policy question for multinational trade negotiations of the GATT system. If negotiators fail to appreciate this voluntary effort, it will be politically difficult for the Korean government to initiate such a program in the future' (ibid.: 88). Therefore, as Table 4.7 illustrates, the *chaebol*s had effectively begun to dominate business in Korea and had also become less dependent on the government (Shibusawa, Ahmad and Bridges 1992: 69).

These reforms were given even greater emphasis by President Roh Tae-Woo's government (1987–February 1993). However, there was considerable

Table 4.6 Major *chaebols*' possession of stocks of commercial banks (as of 31 December 1988)

(Unit: Percentages)

Chaebol	*Korea First Bank*	*Hanil Bank*	*Korea Commercial Bank*	*Cho Heung Bank*	*Seoul-Trust Bank*	*HanMi Bank*	*Other local banks*
Samsung	5.3	4.9 (A)	11.0	7.4	2.7	0.7	DaeGu (11.0)
Hyundai	2.2	1.9	4.5				KangWon (23.6)
Lucky-GoldStar	3.2 (A)	3.9	1.0				
Daewoo						8.6	ChungBuk (1.7); ChunBuk (1.6)
Sunkyung							KyungKi (4.5)
Ssang Young				1.7 (A)			
Hankook Whayak		4.3					ChungChong (11.3)
Hanjin		4.6					KyungKi (11.2)
Hyosung							KyungNam (4.1)
Daelim		3.7 (A)					
Shin Dong (A)	6.5	1.1		6.5	6.3		KyungKi (13.2)
Hanil Hyupsung							KyungNam (1.2)
Dong-Ah					4.2		
Tae Kwang				8.7	1.1		
DongKuk					1.5 (A)		Pusan (4.4)

Sources: *Mail Kyung Jae ShinMun* [Daily Economic Newspaper] 19 May 1989, and *HanKook Chaebol Study* [A Study of Korean Chaebols] (Cho 1990: 149).

Note: A: Figures apply to end of 1991.

opposition both within and outside the government to some of the liberalisation proposals,[15] which slowed reform progress during the period (ibid.: 70).

Cho and Hellmann explained this as follows:

> At the more advanced stages of economic development, the role of the GLIO diminishes. Innovation superseded learning and adaptation as an engine of growth. Because the environment must support individual entrepreneurship and innovation, an internal organization may fail to provide the correct incentives for innovation, especially

Table 4.7 *Chaebols*' share in total sales volume in Korea (selected years 1985–94) (percentages)

	1985	*1987*	*1989*	*1990*	*1992*	*1994*
5 largest *chaebols*	12.4	12.6	12.6	10.2	11.3	11.2
10 largest *chaebols*	18.9	18.9	18.2	15.3	16.2	16.1
20 largest *chaebols*	25.7	25.3	24.4	21.3	22.2	22.3
50 largest *chaebols*	34.8	33.5	33.0	29.2	30.6	30.9

Source: *1996 Nyn 30 Dae KieupGipdan* [Top 30 Largest Business Groups in 1996] (Choi 1996, Table 3.2).

> since (intellectual) property rights within the internal organization tend to be vaguely defined. The consensus on the direction of a development strategy is likely to erode, and different fractions of society become more likely to push for their own socio-economic agendas, which put less emphasis on national economic growth.
>
> (1993: 29)

This argument explains the diminished importance of the GLIO as Korean economies developed. The more subtle issue is whether only the role of government (GL) or the entire institutional approach (IO) diminished in importance. Cho and Hellmann argue that the GL became dispensable at higher stages of economic development, whereas the IO adapted to the changes in the economic environment with the expansion of PLIOs. But, after institutional arrangements had been developed and the private sector had the expertise to organise economic activity, there were many options to reduce the government's involvement. One clear advantage of private institutions was that control over economic activity was better insulated from political misuse. The transfer of control to the PLIOs had the advantage of maintaining some of the institutional advantages of internal organisation, while at the same time relaxing the government's grip on the economy.

Although the strength of this relationship was clearly weakened during this period, the relationship between the government and business in Korea remained fundamentally unchanged. The government still had a whip hand in relations with business. Thus, it was still the fact that a key to the *chaebols*' success during this period was their usefulness to the government as an instrument for economic development (Steers, Shin and Ungson 1989: 42).

Several events show that the government still had a senior position in the relations between the two. One was the so-called Kukje group case. Kukje was Korea's seventh-largest *chaebol* in 1985 when it was forced into bankruptcy by the government. At that time, the Kukje group had a diversified portfolio of profitable companies. It had 38,000 employees and an annual turnover of US$ 1.5 billion. However, like most other *chaebols*, it was highly leveraged and had borrowed considerably from the government. Despite official denial' it was widely believed that the Kukje group was forced into bankruptcy by the government because Mr Yang Chung-Mo, ex-chairman of the group, failed to contribute generously to government projects. According to Mr Yang, he did not provide sufficient support to government-sanctioned causes such as the *Saemaul Undong* (New Village Movement)[16] and the Ilhae Foundation,[17] although it was a common practice in Korea for companies to back government-supported causes by donating money (in fact, it was seen as a form of quasi-tax) to the government. The Kukje group donated only US$ 400,000 to *Saemaul Undong* when other large *chaebol*s gave more than US$ 1 million annually (ibid.: 43). Moreover, Yang refused to participate in a US$ 40 million campaign drive for the Ilhae Foundation. It has been widely charged that these payments represent bribes to the government in exchange for government support of the companies.

The second case was KS (Korea Shipping) Line, the largest shipping cargo company in Korea in the late 1970s and early 1980s. According to Mr Yun Suk-Min, ex-chairman of KS, the company was also forced into bankruptcy by the government in 1987 for reasons similar to the Kukje case (*Wol Gan Chosun* (magazine) January 1994). Mr Yun was the deputy leader of the People's Party, then the second opposition party. In this role he criticised many policies and even leading politicians of the Democratic Justice Party, which was then the ruling party led by President Chun. Therefore, the government could be expected to retaliate.

The message was clear: if *chaebol*s maintained a good relationship with the government by supporting the ruling political party and making donations to government-supported causes, they would prosper. But failure to comply could – and sometimes did – lead to a termination of financing and immediate bankruptcy.

As a way to maintain good relationships with the government, companies have often hired those who were retiring from high government positions as top executives, as illustrated in Table 4.8. This practice has benefited both

Table 4.8 Previous occupations of top managers*

	Politicians	*Bureaucrats*	*Military*	*Financiers*
Samsung	—	1	2	1
Hyundai	–	2	3	2
Lucky-GoldStar	2	4	2	2
Daewoo	—	5	6	12
Sunkyung	—	1	—	1
Ssangyong	—	4	2	1

Source: *Wol Gan Sin Dong A* [Monthly Magazine Sin Dong A] (August 1986: 403).
Note: * All of the above managers were hired with a title of executive vice-president or higher.

parties because the government retirement age has historically been 55, while the corporate retirement age among top executives has been somewhat more flexible. This practice has been effective for firms because a relatively high number of government officials want to work after their retirement due to increased average life expectancy in Korea. The average life expectancy in Korea increased from 53 years in 1960 to 72 years in 1990 (World Bank 1993a).

The companies also hired from the military (although this could be considered government hiring in view of the close military–government ties under Presidents Park and Chun) and from financial institutions. It is important to remember that many other countries, including the United States and France, have had similar practices of hiring key government employees from the military and the government itself, although possibly not to the same extent as Korea.

Another tactic was for companies to try to establish family relations with top government officials through marriage among their sons and daughters. For instance, the eldest son of the chairman of the Sunkyung Group got married to a daughter of former president Roh Tae-Woo. A daughter of chairman of Dongbang Yuryang also married a son of Mr Roh. A daughter of ex-chairman of POSCO (Pohang Iron and Steel Company Ltd.), the second-largest steel complex in the world, married the eldest son of former President Chun Doo-Hwan.

(iv) The relationship between government and business since 1993

Since 1993, the relationship between the two has remained unchanged. However, there has been an intimation of changes to come in the relationship between government and business. As an example, after the launch of the Seventh Republic, Kim Young-Sam, newly elected President, announced that he would not accept any money from firms for any purposes. In addition, Yang Chung-Mo, ex-chairman of Kukje, and Yun Suk-Min, ex-chairman of KS brought their cases to the high court. It was concluded in the Kukje case by the Constitutional Court on 27 July 1993 that the government had acted unconstitutionally. It can be said that in Korea, big-business firms have started to have more influence in their relations with the government. However, the government is and will continue to be firmly in the driver's seat for the time being. In fact, the then Minister of Finance, Hong Jae-Sung at an interview in July 1993, said that government could not control the financial market directly, as it used to, due to the size of the Korean economy, but that the government still wielded more power than firms or banks in the relationship. He also said that he agreed with the principle of putting limits on an individual's or an organisation's holding of stocks of the second group of financing and banking businesses[18] because there was a limit on individual's or an organisation's possession of stocks of the commercial banks (at maximum 8 per cent), which are classified as belonging to the first group of financing and banking business (*Ju Gan Sisa Journal* 8 July 1993: 62–3).

A good example proving that the government still has a strong position in its relations with business in Korea is the Hyundai case. Founder and former chairman of Hyundai, Mr Chung Ju-Yong ran for the 1992 Presidential Election as a strong competitor to newly elected President Kim. During the campaign, Mr Chung criticised Mr Kim personally. Because of this, relations between Mr Kim and Mr Chung soured. After the election, the government, led by newly elected President Kim, allegedly put visible and invisible pressures on Hyundai. From just after the presidential election to the beginning of 1995, the Korean government is rumoured to have pressurised commercial banks directly and indirectly to stop offering bank loans to any companies belonging to the Hyundai group. In December 1994 the government turned down an application by the Hyundai Group to set up a steel business.

4.1.2 The relationship between government and business in Japan

In Japan, the relationship between government and business has typically been one of mutual consensus among relative equals on policy decisions (Cho and Hallmann 1993: 10; Steers, Shin and Ungson 1989: 44). In other words, the relationship between government (especially the Ministry of International Trade and Industry, or MITI) and business in Japan has not been completely equal but has been at least a little less one-sided than in Korea (Steers, Shin and Ungson 1989: 47). One of many examples that demonstrates the relations between government and business in Japan is Japan's financial regulation. The record of business–government interaction as covered in Japanese financial regulation casts serious doubt on the common characterisation of Japanese economic performance as government-led (Rosenbluth 1992). As Rosenbluth (ibid.: 33) insists, 'rather than utilizing the financial system as an instrument of state policy, the Japanese government has in large part responded to the interests of the well-organized and politically influential banks in formulating financial regulation'.

Ito explains the relationship between the two in Japan as follows:

> The relationship between business and government in Japan looks very different from that in the United States. The conventional wisdom emphasizes a cooperative relationship between the government and big business: Government and business jointly choose 'target' ('sunrise') industries; they then protect these industries with high tariffs and import quotas and nurture them with preferential loans and other resources. After a target industry reaches maturity, government and business begin a drive to increase exports. MITI (the Ministry of International Trade and Industry) also arranges depression

> cartels for industries experiencing temporary declines in demand, and coordinates the orderly exit of firms from structurally declining ('sunset') industries.
>
> (1992: 178)

One of the main reasons for differences between Korea and Japan in the relationship between government and business has come from the nature of financing. In Japan, as in Korea, the government has relied on financial and monetary means to guide and control private activities. Most borrowing for new ventures in Korea has come from the government or from commercial banks which have been influenced strongly by the government, while in Japan far more borrowing has come from group-connected banks, which has given Japanese firms greater freedom than their Korean counterparts (Steers, Shin and Ungson 1989: 47). In Japan, in fact, most conglomerates have included banks as part of their organisational structure and have been thus assured of reliable sources of credit, as illustrated in Table 4.9. Japanese banks and firms, in the pre-war period, were interrelated through direct ownership within larger family groups (*zaibatsus*), and in the post-war period, through substantial ownership links within *keiretsus*, somewhat different forms of the old industrial conglomerates (*zaibatsus*).

In addition, in the post-war period, industrial firms clustered around main banks, with which they had a close monitoring relationship. Therefore, the Japanese government could not guide and control private business activities as directly as the Korean government did through financial and monetary means.

In general, the relationship between government and business in Japan has remained unchanged since the end of the Second World War, as illustrated in Figure 4.1(B). Before that, the relationship between business and government in Japan was quite different. Before the war the government intervened in private business activities more directly and extensively. This was possible because there were many state-owned firms during the period. In fact, the Japanese government began intervening heavily in credit allocation in the 1930s for the promotion of the HCI, at first through the Industrial Bank of Japan (IBJ), and then later, through the private commercial banks. During the pre-war period, especially during wartime, the government controlled firms by intervening in credit allocation in this manner.

However, in post-war Japan, the role of the government was less extensive and more indirect: the private sector was able to make mostly independent decisions, but firms co-operated with each other and generally with the government (Cho and Hellmann 1993: 13). In other words, after the Second World War, the role of the government in Japan diminished as the country industrialised and as the private sector expanded its own internal market. 'The heavy government intervention in credit allocation during wartime influenced the postwar relationship between the government and banks,

Table 4.9 Japanese conglomerates' possession of stocks of commercial banks (as of 30 September 1992)

Dai-ichi Kangyo Bank (1)	*Sakura Bank (3)*	*Mitsubishi Bank (4)*	*Fuji Bank (5)*	*Sumitomo Bank (6)*	*Sanwa Bank (7)*
Asahi Life Ins. (4.4%)	Nippon Life Ins. (3.5%)	Meiji Life Ins. (5.8%)	Yasuda Life Ins. (4.9%)	Sumitomo Life Ins. (6.0%)	Nippon Life Ins. (4.5%)
Nippon Life Ins. (3.7%)	Mitsui Life Ins. (3.5%)	Tokio M. & F. Ins. (4.3%)	Dai-ichi Life Ins. (2.9%)	Nippon Life Ins. (4.4%)	Meiji Life Ins. (3.2%)
Dai-ichi Life Ins. (2.8%)	Taiyo Life Ins. (3.5%)	Dai-ichi Life Ins. (3.5%)	Nippon Life Ins. (2.9%)	Matsushita Electric. Ind. (2.7%)	Daido Life Ins. (3.0%)
LTCB (2) (2.8%)	Dai-ichi Life Ins. (2.9%)	Nippon Life Ins. (3.1%)	Yasuda F. & M. Ins. (2.3%)	Sumitomo Trust (2.4%)	Dai-ichi Life Ins. (2.4%)
Meiji Life Ins. (1.8%)	Toyota Motor (2.5%)	Mitsubishi Heavy Ind. (3.0%)	Nippon Steel (2.1%)	Sanyo Electric. (2.0%)	Toyota Motor (2.2%)
Nippon Credit Bank (1.3%)	Toshiba Corp. (1.4%)	Mitsubishi Trust (1.8%)	Nissan Motor (1.8%)	Dai-ichi Life Ins. (1.9%)	Nippon Steel (1.7%)
Fukoku Life Ins. (1.3%)	Mitsui Trust (1.2%)	Taiyo Life Ins. (1.7%)	NKK Corp. (1.7%)	Sumitomo Corp. (1.8%)	Teijin Ltd. (1.6%)
Kawasaki Steel (1.2%)	Mitsui M. & F. Ins. (1.0%)	Mitsubishi Corp. (1.7%)	Nihon Cement (1.6%)	Kajima Corp. (1.7%)	Toyo Trust (1.4%)
	Meiji Life Ins. (1.0%)	Nippon Steel (1.6%)	Taisei Corp. (1.6%)		Hitachi Ltd. (1.4%)
	Mitsui Fudosan (0.9%)		Yasuda Trust (1.5%)		Nippon Express (1.4%)

Source: *Toyo Keizai* (spring 1993).

Notes: 1: Largest city bank. 2: Long-term Credit Bank of Japan. 3: Nucleus of Mitsui group. 4: Fifth-ranking city bank in terms of fund volume. Nucleus of Mitsubishi group. 5: Fourth-ranking among city banks in fund volume. Nucleus of Fuyo group of companies affiliated with old Yasuda *zaibatsu*. 6: Third-ranking city bank in fund volume. Nucleus of Sumitomo group. 7: Sixth-ranking city bank in fund volume. Nucleus of Sanwa group of Sansuikai.

which continued close consultation and maintained a cooperative relationship' (ibid.: 11). After the war, in fact, the Japanese government could have felt less compelled to intervene by directing credit explicitly because the institutional environment could overcome market imperfections to some extent. After 1945, most of the industrial firms were privately owned, except for some key strategic industries. The private firms, especially large industrial conglomerates (*keiretsu*) have assumed dominant positions in the Japanese economy, as Table 4.10 shows.

Table 4.10 Economic size of six enterprise groups (1984 and 1987)

	No. of member firms A(1)	B(2)	*Average inter-locking shares (%)*	*Average intra-group loans (%)*	*Total employees (3)*	*Total assets (¥ billion) (3)*	*Sales (¥ billion) (2)*	*GNP (¥ billion)*
FY 1984								305,896
Mitsui	24	20	17.71	19.68	245,205	46,717	27,824	
Mitsubishi	28	24	24.91	23.35	241,533	60,997	28,616	
Sumitomo	21	17	24.87	27.15	154,964	54,472	19,587	
Fuyo	29	25	15.72	17.21	336,954	59,388	28,315	
Sanwa	42	35	16.56	19.57	400,191	63,106	29,334	
Ikkan	47	42	13.67	12.01	496,403	66,336	45,860	
FY 1987								344,653
Mitsui	24	20	17.10	21.94	238,447	59,525	27,405	
Mitsubishi	29	25	27.80	20.17	241,846	85,524	24,088	
Sumitomo	20	16	24.22	24.53	153,202	79,197	20,011	
Fuyo	29	25	15.16	18.20	322,798	77,929	26,861	
Sanwa	44	41	16.47	18.51	377,622	88,148	31,080	
Ikkan	47	42	12.49	11.18	466,250	86,345	48,653	

Sources: Based on (Ito 1992: 181, Table 7.1): *Major Statistics of Korean Economy* (National Statistical Office, Republic of Korea 1993).
Notes: A(1): Total. B(2): Without banks and insurance companies. 3: With banks and insurance companies.

Even small- and medium-sized firms that did not belong to a *keiretsu* then had 'main banks' which, as in Germany, held equity participation and representation on the firms' boards. Small firms typically had a subcontracting relation with large firms that gave a certain control over their business.

One study of government–business relations in Japan concluded that 'big business is predominant and unrivalled as an influence in Japanese politics. Wishes of big business are tantamount to commands, and the government does not dare to take them lightly' (Yanaga 1968). Cho and Hellmann (1993: 13) noted that 'for post-war Japan, we witnessed a looser affiliation between the government, banks, and large industrial firms. The GLIO was complemented by strong private internal organizations that linked the banks and industrial firms more closely.'

Generally, then, it can be said that relations between government and business in Japan have been based on mutual consensus between relative equals on policy decisions, despite the fact that Japanese firms still depend on the government to conduct trade relations with other countries, manage exchange rates and pressure international organisations such as the OECD, IMF, GATT, and so on. The reason for this is that, particularly after the Second World War, businessmen were not obliged to follow 'administrative guidance' as they once were, as they could obtain funds from related banks or by issuing stock and getting technologies from overseas partners or creating their own through R&D. For a time, before they had the capability to obtain the necessary funds from banks, or to issue stock and get technologies from overseas partners or to create their own through R&D, those in business were gradually altering the power relationship with the government by financing politicians who then had to repay their debts. By doing so, they could seek favours for business in general (*zaikai*) or in specific sectors. In Japan, to be a politician demands a considerable income. Thus, many politicians had debts.

As Woronoff (1992: 44–5) put it, the businessmen also provided more of the input for national policy in areas in which they were involved. Through leading business organisations like the Federation of Economic Organisations (Keidanren) or the Federation of Employers Associations (Nikkeiren), they exercised influence on government policy and programmes, even to the extent of bringing about administrative reform. In fact, the ministries and agencies wanted to discuss upcoming operations with businessmen because they knew far more about actual business operations than any bureaucrats.

4.1.3 The general relationship between government and business in Taiwan

In Taiwan, a period of primary import substitution that lasted until about 1960 was followed by a phase of export-oriented industrialisation lasting until the early to mid-1970s. After this period came liberalisation. Throughout the whole period, the Taiwanese government continued to intervene in private business. This was against the familiar neoclassical assumption that an increase in outward orientation goes with a reduction in government intervention. In this way, the Taiwanese government

> was the central actor in guiding and coordinating the whole process of industrialization, by setting up the needed infrastructure, attracting foreign capital, deciding the priorities for strategic investment, and imposing its conditions when necessary (for instance, the first attempt to initiate a Taiwanese automobile industry failed when

the government refused to accept the conditions required by Toyota).

(Castells 1992: 44)

However, the relationship between government and business in Taiwan had been less superior–subordinate than that in Korea but more superior–subordinate than in Japan, as illustrated in Figure 4.1. In other words, 'the Taiwanese government has relied more on arm's-length incentives to steer private firms, often used public enterprises or public laboratories to undertake big pushes in new fields' (Wade 1990: 321).

In Taiwan, government planning only provided guidelines for the private sector, which could go its own way 'but tended to weight the possible advantages of conforming with official goals and policies. This made the exercise somewhat more effective and purposeful than in Japan if hardly constraining' (Woronoff 1992: 68). There were several reasons for such a relationship between government and business in Taiwan.

First, the Taiwanese government, dominated by the Kuomintang, had less legitimacy than the Korean government. Taiwan's Ministry of Economic Affairs had less power than its Korean counterparts, the Economic Planning Board and the Ministry of Trade and Industry, both of which had enough power to determine which companies received which loans at what rate. This was due mainly, to the Korean government's greater responsibility for the budget and greater influence over monetary policy (tight control of banking) (Wade 1990: 322–3). In addition, the Bank of Korea, the central bank, had been subordinate to the Ministry of Finance, unlike the situation in Taiwan.

In other words, the Korean government had a more centralised management structure, one which assigned pre-eminence in industrial policy to the Ministry of Trade and Industry and the Economic Planning Board. Top officials of these government agencies have been able to exercise broad control from a single position, acting through command hierarchies and relating to business groups as leaders rather than equals.

The Taiwanese government, on the other hand, was less centralised, with power over industrial policy issues dispersed amongst more ministries and agencies. Officials had narrower scope for the exercise of their authority and used it more circumspectly in their dealings with private firms. They relied more on public enterprises for initiatives in high-entry-barrier sectors.

Second, the Taiwanese government had, in fact, exerted some influence on the market through the mass of publicly owned assets and organisations (ibid.: 323), although small- and medium-sized enterprises played a more important role in the Taiwanese economy, as illustrated in Figure 4.2. In fact, as of 1992, enterprises employing fewer than 50 people contributed 94 per cent of Taiwan's total production of goods and services (Shibusawa, Ahmad and Bridges 1992: 74). Through this mass of publicly owned assets

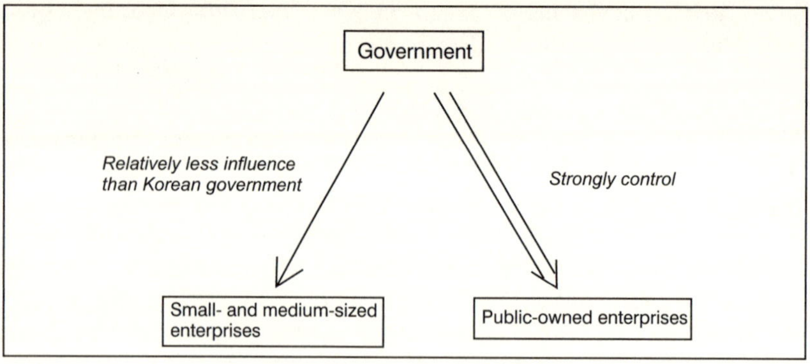

Figure 4.2 The relationship between government and business in Taiwan.

and organisations, the Taiwanese government was able to guide and influence the market no less than the Korean government, but in a different way. It relied less on direct steerage of private investment decisions and more on public investments to induce a downstream response. It also made the cost of entry for small businesses very low, in order to prevent greater resentment by the islanders against the Kuomintang mainlanders. But the difference with Korea is one of degree rather than kind. Because of differences in the economic development strategies of Korea and Taiwan, the Taiwanese government did not control private firms (most of them were small- and medium-sized firms) as much as the Korean government.

The difference in economic development strategy between Korea and Taiwan was due in large part to political factors. The Taiwanese government deliberately encouraged smaller units of production because of political factors. In Taiwan, 3 million mainlanders held political power and regulated the economic activities of 10 million Taiwanese. Thus, the government deliberately attempted to prevent alternative power-centres, such as the *chaebol*s in Korea, from emerging in order to avoid upsetting its power base. Due to this government strategy, the size of firms in Taiwan was smaller than that in Korea.

Third, the Taiwanese government was never much involved in private business. The state sector is much bigger in Taiwan than in Korea. In 1976 public enterprises accounted for 22 per cent of Taiwan's GDP but for only 9 per cent in Korea. The reasons why the government did not involve itself in private business were as follows. The first reason is that it would have increased tension between the islanders and the Kuomintang mainlanders. The second one is that it would have gone against advice from the United States, a major aid donor. The final reason is that excessive state involvement might discourage sorely needed foreign investment (Woronoff 1992: 70). In short, excessive state involvement was against the Kuomintang's interests as well as its ideology. Moreover, the political leaders in Taiwan were preoccupied with other fundamental issues of domestic and foreign policy, defence and the threat from the People's Republic of China.

Fourth, the government owned major banks, as did the Korean government, and industrial corporations in Taiwan, so that it could control business more strongly than the Japanese government did.

In Taiwan, financial control and loan allocation have been crucial to economic growth, much as in Korea, but the form has been different. The Taiwanese government has tended to rely on monetary rather than fiscal policies – tax breaks and high-depreciation allowances rather than outright loans – to encourage investment in particular sectors. Moreover, most Taiwanese government loans go to state-owned enterprises rather than to big businesses.

The Taiwanese government has also used credit, taxes, domestic content requirements, trade policy and direct foreign investment controls to steer private investment decisions, though less forcibly overall and with less reliance on selective credit.

Recently, the government has tried gradually to reduce its own direct ownership of major banks and enterprises. In May 1989 the Cabinet approved plans for privatising or partially privatising 15 major banks and industrial corporations.

'In short, the government's economic planning agencies, under President Lee, are trying to aid industrial restructuring by moving towards a coordinating role, rather than following the directive, at times autocratic, style that prevailed previously' (Shibusawa, Ahmad and Bridges 1992: 75).

However, the Taiwanese government still intervenes in private business activities, less so than the government of Korea but more so than that of Japan.

4.2 The relationship between government and business in Korea in the decision-making on FDI by Korean consumer electronics firms in the EU

As discussed previously, the Korean government has habitually taken the initiative in relations with business, although in general the role of government has been gradually reduced in recent years. Strangely, the Korean government reacted in a different way to Korean consumer electronics firms investing in the EU. In more detail, unlike the existing relationship between the government and business in Korea, according to the results of a questionnaire and a series of interviews with senior Korean businessmen and a semi-governmental official[19] in the EU, firms first recognised the necessity of FDI in the EU and then asked the government to take necessary follow-up actions[20] when they were considering investing in production in the EU.

According to the results of my questionnaire, all 16 companies which replied gave the same response.[21] In more detail, six of these companies[22] said that their chairmen or chief executives had initiated the idea of making FDI in the EU, mainly in order to avoid discriminatory restrictions such as tariffs and quotas on Korean exports to the EU, as well as to implement a globalisation strategy. Seven companies[23] replied that managers felt the necessity for FDI in the EU. Three companies[24] said that both chairman/chief executives and top- and middle-level managers felt the necessity for FDI in the EU.

If, in general, the Korean government was reducing its intervention in business, why did it respond so positively to Korean consumer electronics firms considering and making FDI in the EU?

First of all, the Korean government acknowledged that, in countries such as their own, the economy is largely influenced by local firms in global competition. This in turn affects national competitiveness and living standards. One of the best ways to measure the success of Korean firms in global competition is in terms of export rates to GNP. In fact, Korean export rate to GNP had been in the range of about 30 per cent to 40 per cent between 1973 and 1994, which had been quite high in comparison to that of other countries. In particular, the success of the consumer electronics industry in global competition, one of six strategic industries that the Korean government chose to focus on,[25] has been very important for national competitiveness of Korea.

There are two useful definitions of national competitiveness. Porter explains national competitiveness as follows:

> The only meaningful concept of competitiveness at the national level is productivity. The principal goal of a nation is to produce a high and rising standard of living for its citizens . . . A nation's standard of living depends on the capacity of its companies to achieve high levels of productivity – and to increase productivity over time.
>
> (1990a: 84)

Krugman (1994: 34) insists that 'the growth rate of living standards essentially equals the growth rate of domestic productivity–not productivity relative to competitors, but simply domestic productivity.'

These two definitions apply well to the Korean case. In addition, through its experiences during 1979 and 1980, the Korean government acknowledged that too much intervention in markets had resulted in economic failures leading to the decrease of the exports, causing an overcapacity in many target sectors, such as HCI, a reduction in effective competition, concentration of economic power, a waste of resources and many serious sectoral imbalances and structural problems of the economy (Jung 1989: 24; Sakong 1993: 72). In other words, the government acknowledged the success of Korean firms in global competition to be the most important factor for the living standard of the people. Thus, the Korean government took four major steps to correct these problems in the early 1980s. These steps were: '(1) more reliance on price and market mechanisms; (2) strong emphasis on price stability; (3) efforts to open the domestic market to foreign competition and (4) promotion of small and medium companies' (Jung 1989: 24).

After this change in the government's attitude to its own role in economic development, European integration, which started in 1985 with the Single Market Programme, was the first opportunity for both the government and business in Korea to respond to changing world market conditions. This changed the government–business relationship. Korean exports to the EU had been blocked due to non-trade barriers, while the EU also emerged as the largest single market in the world. Under these conditions it was quite natural that Korean firms, not the Korean government, should start showing great interest in the situation in the EU and taking the necessary actions.

This view was confirmed by the interviews with senior managers of nine Korean manufacturing firms which had invested in the EU. All of them agreed that they felt the necessity for FDI in the EU first, and that the Korean government just took follow-up actions at the request of the firms.

Mr J.R. Suh, director of the Korea Trade Centre, London, which is a semi-governmental agency, confirmed that firms took an initiating role in decision-making regarding FDI in the EU and that the government simply took a supportive role. He added that the Korean government has made great efforts in attracting foreign investments to Korea, while it has made less effort in encouraging Korean firms to make foreign direct investment.

In addition, one thing Korean consumer electronics manufacturers who made FDI in the EU had in common was that most of them were *chaebol*s (in fact, only three of 24 companies were not *chaebol*s). These companies belonging to a *chaebol* also had trading companies. This means they could get necessary information for making FDI in the EU from the branch offices of their trading companies. According to the results of the questionnaire, the major motives for their direct investment in the EU were to carry out globalisation strategies and also their concern about a rise in protectionism in the EU.[26]

Another reason for the reduced intervention of the Korean government in business is simply that political leaders in Korea were extremely preoccupied with domestic political problems and with preparations for the 1988 Seoul Olympic Games. From 1985, when the plan of the Single European market was announced, the Korean government had less time and effort to spare for economic issues. For several months before the Presidential Election in December 1987, in particular, the government of President Chun Doo-Hwan was forced to take seriously demands, not only for a fairer distribution of the benefits of economic development but also for a greater share of the political cake at a time when there were popular uprisings involving many middle-class Koreans against the government (Shibusawa, Ahmad and Bridges 1992: 69). The nation-wide character and the broader social involvement of these protests forced Roh, who was nominated as the presidential candidate of the ruling Democratic Justice Party, and ultimately President Chun, to promulgate a new and more democratic constitution in June 1987. Mr Roh announced a 'democratisation package' that met almost all of the opposition's demands, including direct presidential elections because he recognised that martial law would have caused bloodshed and chaos, and almost certainly would have meant Seoul's loss of the 1988 Olympics, which meant so much to all Koreans (ibid.). At the Presidential Election in December 1987, Chun's chosen successor, Roh Tae-Woo, was elected with less than 40 per cent of the popular vote because his major opponents, Kim Young-Sam of the Reunification Democratic Party (RDP) and Kim Dae-Jung of the Party for Peace and Democracy (PPD), failed to unite their parties. In the General Election for the National Assembly held in April 1988, the Democratic Justice Party (DJP), the ruling party, failed to win a majority against the three major opposition parties (PPD, RPD, and the New Democratic Republican Party or NDRP) for the first time in Korea's history (Youn 1989). In sum, then, Korean political leaders who took power in 1980 by a military *coup d'état*, were extremely preoccupied with domestic political problems.

Third, US protectionist pressures against Korea also gave the Korean government a chance to recognise the importance of maintaining a dynamic economy. In fact, US protectionist pressures came as a shock to the Korean government because the United States has been the most important trading partner for Korea, and both countries have, since the Korean War, maintained a very strong political and military alliance. Since the mid-1980s, US interest groups and politicians have called for several actions such as protectionist measures against Korean exports, greater opening of the Korean market, adherence to international copyright and patent conventions, and most recently, the realignment of the war.[27] According to these demands, 'the United States government used the threat of punitive action under the notorious Super 301 clause of the 1988 Omnibus Trade Act to extract further concessions from Korea in both 1989 and 1990' (Shibusawa, Ahmad and Bridges 1992: 72).

Due to the experiences in the late 1970s, the Korean government realised the need to reduce its involvement in the private sector in order to facilitate a dynamic export-oriented economy.

Fourth, in practice, it was possible for Korean consumer electronics firms to make FDI in the EU by themselves, largely because Korea's economy had become too large and complex to be subjected to central control (Woronoff 1992: 117). The Korean government recognised that government planning and intervention were warranted when an economy was small and in a primitive stage (Porter 1990a: 86). However, 'the more command elements, including targeting, had to be discarded and businessmen were urged to follow the market signals more than the plan' (Woronoff 1992: 117).

4.3 Conclusion

The relationship between the government and business in Korea has been interventionist in nature compared to that in Japan. The Korean government's role has been considerably more extensive than merely setting the broad rules of the game and influencing the market indirectly through economic policy, although government influence over private business activities has undoubtedly been weakened to some degree since the beginning of the 1980s. Indeed, it has been argued that continuing effective government control of the banking sector has resulted in a still subordinate position for most of the private sector.

The general relationship between government and business in Japan has not been completely balanced either, but it has typically been one of mutual consensus between relative equals on policy decisions. In fact, in post-war Japan, the role of government has been more indirect, and the private sector has been able to make mostly independent decisions in co-operation with itself and the government. The Korean and Japanese governments have both each used financial and monetary means to guide and control private activities, particularly since the Korean government owned commercial banks until the early 1980s. This meant the government could exert control over firms. In both countries the role of government diminished after industrialisation and as the private sector expanded its own internal market. This evolutionary process happened earlier and has gone further in Japan than in Korea.

In comparison with the Korean government, the Taiwanese government has also relied less on direct control of private investment decisions and more on public investments to induce downstream responses. It has also made the cost of entry for small businesses very low. However, the difference between the Korean and Taiwanese approach is one of degree rather than kind. The Taiwanese government, too, has used credit, taxes, domestic content requirements, trade policy and direct foreign investment controls to influence private investment decisions, though less forcibly overall and with less reliance on selective credit. The Taiwanese government has relied both on monetary and fiscal policies. However, the Taiwanese government has tended to rely on monetary rather than fiscal policies – tax breaks and high-depreciation allowances rather than outright loans – to encourage investment in particular sectors. And in Taiwan too, the role of the government has been weakened.

In short, the relations between government and business in these three countries were characterised by the coexistence of authoritarianism and capitalism rather than the domination of capitalism by authoritarianism. Such government–business relations are nevertheless a world-wide issue, not just an East Asian one. Besides these three East Asian countries, the governments of some Latin American countries, for example, have similar government–business relations. However, in more detail, there are some differences in the government–business relations between East Asian countries, such as Korea and Taiwan, and Latin American countries (White and Wade 1988: 9). The differences, in brief, are that 'state intervention in East Asia has been both stronger and more selective than that in Latin America, not only at the national boundary but also in key parts of domestic industry' (ibid.).

The Korean government responded very differently to Korean consumer electronics firms considering investing in the EU. In the decision-making process of making FDI in the EU, according to the results of my interviews and questionnaire, Korean firms did almost everything they could to achieve FDI in the EU, acting completely independently from government. In fact, as discussed earlier, the firms felt the necessity of FDI in the EU based on the information that they obtained and decided to undertake FDI in the EU. Only after this did the firms ask the Korean government and the Bank of Korea to take the necessary follow-up action, such as increasing the maximum amount of FDI which firms could make without government approval. This was possible for the following reasons. First, the Korean government acknowledged that national competitiveness, which largely affects the standard of living of the people, is decided by the success of firms in global competition. This is particularly so in countries such as Korea with its economy largely influenced by external markets. This is in line with Porter's explanation. Porter (1990a, 1990b) explained that a limited degree of state intervention within a neoclassical framework stressing unfettered market competition is a determinant of economic success among nations. However, Porter's idea does not fully explain the economic development which Korea has achieved. Second, political leaders in Korea were extremely preoccupied with domestic political problems and the preparation for the 1988 Seoul Olympic Games.

Third, US protectionist pressures against Korea, particularly in the form of Super 301, made the Korean government recognise the importance of maintaining a dynamic economy, because Koreans think the United States (which they previously regarded as sympathetic to Korean interests) is unfair in at least three ways:

> [T]he United States reacts forcefully to others' protectionism, while simultaneously practicing protectionism against politically weak countries; the United States treats Korea like an advanced industrial country, such as Japan. Actually, Korea is only a mid-level developing country with a per capita income of US$ 2,300 in 1986. A trade surplus, whether bilateral or overall, is not necessarily a sign either of development or of unfair trading. Continuing the gradual process begun in 1978, Korea is liberalizing imports with renewed vigor despite various difficulties. The United States, however, gives Korea no credit for this laudable unilateral effort. No sooner does

> Korea make a concession than the United States returns with another request and the same protectionist threat.
>
> (Young 1989: 121)

Fourth, Korea's economy has now become so large and complex that it can no longer be subjected to central control. Hence, command elements, including targeting, have had to be discarded and businessmen have been urged to follow market signals rather than a centralised plan (Woronoff 1992: 117). This is illustrated by the failure of HCI policy which was subjected to excessive government intervention. The HCI drive of the 1970s was characterised as an 'industrial targeting approach', in which selected heavy and chemical industries were promoted. This new approach was supposed to upgrade Korea's industrial structure and thereby its export structure for the future, and to offset the negative effects of developed countries' arising protectionist barriers against labour-intensive light manufacturing products – which began after the first oil shock, by exporting higher-value-added HCI products (Sakong 1993: 41). The new pattern of relations between the Korean consumer electronics firms and the Korean government in making FDI in the EU should be understood in the same context.

5

THE IMPACT OF EU POLICIES ON FDI BY KOREAN ELECTRONICS FIRMS IN THE EU MEMBER STATES

During the second half of the 1980s, European countries continued the process of integration by preparing to take four steps, all of which would have marked effects, not just on their own behaviour, but on that of the world as a whole throughout the 1990s and beyond. The first step was the decision to create the Single European Market (SEM) in 1985 among the members of the European Union. The second step was the decision to extend the membership of the European Union to the members of European Free Trade Association (EFTA) through the European Economic Area (EEA) Agreement that was signed in 1992.[1] In addition, before the completion of either of these two processes, the EU has agreed to conclude a new treaty on European Union, which would create an economic, political and monetary union, the terms of which were set out in the Maastricht Treaty. The last element of current European integration, which was very dramatic, was the response to the collapse of the communist regimes in Eastern and Central Europe. These countries' moved towards a market-oriented system with closer integration with the West and the creation of new states out of the former Soviet Union and former Yugoslavia (Mayes 1993: 13).

Of these four elements of, European integration, there has been most progress in the creation of the SEM. The SEM, in particular, has been recognised as a threat to non-EU member countries. Thus, in this chapter, the impact of the SEM on non-EU member countries is the main focus.

Non-EU member states perceived EU integration, especially the start of SEM, as a policy which strengthened protectionism of the EU member states, despite EU efforts to correct such a protectionist image.[2] Oman (1993: 74) stresses that 'the logic driving the Single Market program, and European integration more generally since the mid-1980s, is not one of protectionism *vis-à-vis* the rest of the world.' He continues:

> The logic behind the Single Market program, in particular, is one of strengthening the forces of competition in Europe, including a weak-

> ening or dilution of the rigidifying and growth-retarding effects of nationally entrenched oligopolies, rent seekers and 'distributional cartels', as well as the harmonization of standards and other forms of 'deep' integration. (ibid.)

Nevertheless, it is true that most non-EU member states still perceive EU integration, especially the start of the SEM, as protectionism. This is because the conduct of EU trade policies in recent years has included anti-dumping and anti-subsidy measures, the external tariff together with various preferential arrangements for different non-EU member countries, quantitative restrictions and voluntary export restraints (VERs) with supplier countries, the New Commercial Policy Instrument and the common rules for imports.[3]

According to theories suggested by Mundell (1957), Rock (1973), and Vernon (1966), who insist that the relations between FDI and trade are substitutive, local production through FDI can be regarded as an alternative way to increase market shares in foreign markets if a country strengthens its protectionism against imports from foreign countries. According to this theory, local production of a product in a foreign country should increase if export of the product is reduced in response to the protectionist measures of the country. However, both the volume of exports and of local production of Korean electronics firms in the EU member states have increased, despite the progress of EU integration. But exports might have increased even more in the absence of local production. This fact alone does not disprove substitutability.

There is no absolute answer. It can be said that whether exports are reduced or not depends on (a) market growth and (b) market share by firms. To throw some light on the question, however, this chapter will first examine the general relations between FDI and trade then survey the impact of EU integration on Korean consumer electronics firms operating in the EU. It will go on to assess the general perception by non-EU member states of EU integration and EU policy towards the consumer electronics industry. It will conclude with some general observations on the responses of Korean consumer electronics firms.

Therefore, the main question in this chapter is 'What kind of impact has the start of SEM had on those Korean electronics firms (in particular consumer electronics firms) which were already established in the EU?'

The answer to the question is that each company seeks to maximise business profits, which eventually will bring replacement of exports by FDI in some products.

5.1 The general relations between FDI and trade

Trade and investment can be said to be the two most important aspects of the economic relationship between any two countries or regions and can be thought of as two ways of providing goods to foreign markets for firms seeking competitiveness in international markets.

In general, there is disagreement among experts on whether (or when, or why) FDI is substitutive for trade or complementary and supplementary to trade. It is difficult to generalise about the relation between FDI and trade. On the whole, FDI and trade have no predictable, standard causal relations because there are divergent patterns of experience, such as the period and volume of trade or FDI between the countries concerned, the kind of industry, the features of concerned companies, and types of countries or regions involved. However, if we analyse the relations in more detail, historical experience supplies various explanations of the pattern of relations between FDI and trade (Wilkins 1970, 1974).

Vernon's idea of a product life cycle postulates a hypothesis as to how overseas production (FDI) may eventually substitute for exports of that good (Vernon 1966). Vernon developed his hypothesis on the basis of an analysis of American firms' experiences in the 1960s. American firms were then dominant in foreign investment outflows. In his model, which describes the life of a good from its inception to its eventual standardisation,[4] Vernon looked at the relationship between FDI and trade in a dynamic sense by studying how new products were developed and how declining capital–labour ratios in the production of older goods resulted in change of production place from the innovating country.

Mundell (1957) also explained the relation between FDI and trade as a substitutive one. He demonstrated that trade in goods and international movement of factors could be substitutes because exporting firms chose to penetrate foreign markets through FDI if they had been faced with trade or non-trade barriers. 'This is a much-discussed phenomenon in the context of FDI in developing countries' (Balasubramanyam and Greenaway 1993: 155). However, Markusen insisted that Mundell's explanation of the relationship between FDI and trade focused on an exceptional case. According to him, 'there are sound theoretical reasons for expecting trade in goods and FDI to grow together' (in Balasubramanyam and Greenaway 1993: 159).

Rock (1973) in his study also insisted that FDI would eventually substitute for exports, even though his explanation was different from that of Vernon. Rock rightly analysed the relations between American FDI in developing countries and the trade of these developing countries with the USA in his cross-country study of US FDI in developing countries. He argued that FDI was the last stage of a process that began with exports, which meant that FDI would eventually substitute for exports. According to him, firms start to enter into a foreign market with exports. But the firms may find it necessary to invest there when they face or fear tariff or non-tariff barriers from the foreign market or when their export volume reaches a critical size.[5]

Based on Japanese cases, Kojima (1978a, 1978b) took a different approach to the relations between FDI and trade in his theory. He argues that foreign investments are not a substitute but in fact complement the export of goods, because foreign direct investment causes lower production costs and expedites more exports through a kind of intra-firm trade. Kojima suggested that Japanese FDI originated from established industries in Japan that were losing comparative advantage, but had the potential to achieve comparative advantage in the host country. If an industry in a host country with a potential comparative advantage – which it has not been able so far to achieve because it lacks technology, capital and management skills – were to achieve this advantage through Japanese direct investment, it would develop as a new export industry in the host country.

Roemer (1975) defined the correlation between FDI and trade through observations of the competition between the United States, Great Britain, Germany and Japan. He did not explain the relations between FDI and trade either as substitutive or as complementary. He found a country passing through four stages demarcated in terms of changing mutual strength of its trade and investment. First, its share in world exports of manufactured goods rises; second, its share in trade stabilises and that in world FDI takes off; third, the trade share begins to fall, and in the final stage its share in world FDI also falls. In terms of this cycle, in the mid-1970s, he found that Japan, Germany, the USA and Great Britain were in the first, second, third and fourth stages, respectively.

Encarnation (1992) explains the association between FDI and trade based on his study of the trade relations between the USA and Japan. He explained that strategic policies regarding investment actually complement strategic trade policies. He emphasised that this complementary relation between FDI and trade is clear especially in Japan and other advanced economies, where most trade is intra-company. He adds that two additional market imperfections – large economies of scope and high transaction costs – provide both governments and oligopolists with strong incentives to adopt what he calls 'strategic investment policies'.

Thomsen and Nicolaides (1992) looked for the pattern of relations between FDI and trade in their study of Japanese manufacturing industries in the EU. They try to explain the relations by two approaches: a product-cycle approach and another approach looking at the relations between FDI and trade over the period of time. Through a product-cycle approach, they found that Japanese FDI in the EU substituted for exports to the EU only in a limited sense. According to them, in other words, FDI substituted for exports of certain products only when a multinational enterprise (MNE) faced serious trade barriers. The Japanese companies that engage in FDI make various products which are sold at different prices and made by processes involving varying levels of technology. Thus, these companies put all resources at home into developing new products and technologies, instead of exporting the products when the company faced serious trade and non-trade barriers. In addition, because the process of quality upgrading does not always depend on the development of new products, a firm may at the same time try to upgrade the quality of the products through rationalising its operations both in terms of location and value-added within an existing product range.

Japanese exports to the EU have been most affected by two kinds of restrictions: VERs and anti-dumping duties. Given that a ceiling is imposed on the number of units of the exported product, firms have an incentive to export their higher-value-added models which can command higher prices and profit margins. Lower-value-added models are diverted to other markets. Local production restores the diverted sales of cheaper models rather than replacing exports of more expensive models. In addition, Thomsen and Nicolaides added that:

> as long as exports flow, even though that does not flow freely, firms would have already upgraded their exported products so that subsequent investment and local production are unlikely to result in lower exports of those particular models.
>
> (ibid.: 75)

Looking at the relations between trade and local production over a period of time, Thomsen and Nicolaides found that there was no clear relation between Japanese FDI in the EU and Japanese exports to the EU. What they found from their study was that an increase in one did not lead consistently to a rise or fall in the other. They gave several major reasons why they could not give a clear and unequivocal picture of relations between Japanese FDI in and exports to the EU. The reasons were the absence of a long data series for Japanese cases, the lower level of the overseas production level of Japanese MNEs and severe restrictions against Japanese exports in some sectors. Moreover, the negligible volume of Japanese FDI in the EU before the 1980s meant that the growth rate was easily affected by the experience of one or two firms.

5.2 The general impact of EU integration on FDI and trade

As explained previously, there are four elements of current European integration. When discussing the general impact of EU integration on trade and FDI with non-EU member countries, as Mayes (1993: 13) said, the combined impact of these four elements should be taken into account. However, it is difficult to discuss their combined impact because the three elements other than SEM are still changing. Opinion in most non-EU member countries is that the SEM is essentially a protectionist policy. That alone would justify some estimate of the impact of the SEM on trade and FDI.

In fact, 'the SEM, adopted in 1985, was the first major effort to re-launch economic and political integration in Europe since the enlargement of the Community from six to nine members in the early 1970s' (Pohl and Sorsa 1992: 38). The SEM proposed the reduction or the elimination of most non-tariff barriers to trade within the EU by 1992, through the implementation of 293 measures, as Tables 5.1 and 5.2 illustrate. In more detail, the SEM proposals contained the elimination of technical, physical and fiscal barriers, and the liberalisation of capital and government procurement policies (Bachtler and Clement 1990: 9; Lindert and Kindleberger 1982: 177; Mayes 1993: 13–14; Pohl and Sorsa 1992: 38; Rugman and Verbeke 1991: 33).

Table 5.1 Main features of the single European market programme

The 1992 programme calls for action in six areas	
1 Unified market in goods and services:	• Removal of barriers to trade in financial and other services, and of remaining non-tariff barriers to visible trade. Simplification of customs procedures and elimination of vehicle checks. Harmonisation of technical standards and health/safety regulations, with mutual recognition of certification. Closer alignment of VAT and excise duties.
2 Unified factor market:	• Free movement of capital, with removal of exchange controls. Alignment of savings taxes. Free movement of labour, with mutual recognition of qualifications.
3 Promotion of competition	• Common rules on regulations, takeovers, state assistance to industry, patents and copyrights, company accounting and disclosure of information. Opening of public procurement to competitive tender. Reduced intervention in agriculture.
4 Monetary integration	• Exchange-rate alignment (via ERM). To be followed by adoption of a single currency and creation of a European Central Bank.
5 Social protection	• Adoption of a Social Charter incorporating freedom of movement, fair wages and condition over technological change and company restructuring, protection of children, elderly and disabled people.
6 United response to external challenges	• A common external tariff. Infrastructure projects, especially highspeed rail and road links and integrated telecommunications. Cooperative R&D, especially in microelectronics and information technology. A common energy policy.

Source: 'The Implications of Closer Integration in Europe for Third Countries' (Mayes 1993: 14, Table 1.1).

Table 5.2 Application of major single market measures to third countries

Measures	*Applicability*
Frontier restriction	
• Single-country quotas	• Very difficult to enforce
• Voluntary restriction agreements	• Can continue but lack the sanction of the threat of an effective alternative
• Removal of formalities	• All benefit
Technical barriers	
• Common standards	• Could benefit some members more
• Single certification and testing	• All benefit
Market entry	
• Public procurement	• Members only, unless reciprocity
• Right of establishment	• Some benefit to member states
• Takeover/merger	• Unequal benefits in both directions
Freedom of movement of persons	• Members benefit more
• Other market regulations, e.g. transport, financial services	• Some slight net benefit to members
• Fiscal (VAT, corporation tax, etc.)	• Largely neutral effect
• Financial (state aid, capital markets, etc.)	• Net benefit to third countries

Source: 'The Implications of Closer Integration in Europe for Third Countries' (Mayes 1993: 14, Table 1.2).

The term 'technical barriers' means 'heterogeneous standards related to, for example, health and safety, which often serve to protect domestic producers' (Rugman and Verbeke 1991: 33). These heterogeneous EU standards, in fact, posed many problems. The problems were:

> those of access to the standard-setting process; of obtaining timely information about new decisions; of recognition of non-EU testing or safety procedures; and of potentially erratic enforcement at different EU borders and/or failure of customs officials in one country to accept proof of acceptance by those of another.
>
> (Page 1992: 16)

Page also insists that all the above problems could arise either for legitimate reasons or to hide protectionist aims.

'Physical barriers' primarily means excessive border controls due to administrative requirements such as differences in value-added taxes (VAT) (Rugman and Verbeke 1991: 33). Mayes (1993: 15) and Page (1992: 18) insisted that, in fact, the elimination of physical barriers would benefit EU suppliers for more than non-EU suppliers. According to the arguments of Mayes and Page, goods coming from non-EU member states still have to meet all border requirements and obstacles. And once the fixed costs of this compliance are met, the marginal costs of additional borders before the final destination are small. In addition, for non-EU suppliers, the costs of the delays involved at the frontier may be relatively lower than the costs of shipping long distances to the EU. However, this means that EU producers still have relatively lower prices than that of non-EU producers because they have no frontiers to cross (Mayes 1993: 15).

However, the validity of this argument depends a good deal on the size of firms. Page (1992: 18) argues that small firms of non-EU member states may never reach inside the EU because they cannot compete with the European competitors who enjoy more benefits from the elimination of physical borders.

In terms of fiscal policies and government controls, there will be greater conformity among different systems as a result of (1) 'approximation' and (2) mutual recognition (Rugman and Verbeke 1991: 34). 'Approximation' means, for example, that each EU member state retains the right to fix bands of indirect tax rates (such as excise duties and VAT), instead of requiring these tax rates to be uniform. By doing so, some peaks and troughs in tax rates will be eliminated. 'Mutual recognition' means, for example, that all EU member states recognise financial institutions or professional qualifications accepted in one of the EU member states (ibid.).

In more detail, at the micro-economic level, European countries said they hoped that the measures adopted in order to realise the SEM would bring the following effects:

> a significant reduction in firm costs through economies of scale in production and organization; improved efficiency within companies through industrial reorganization and lower costs rising from a more competitive market; new patterns of competition between industries and improved efficiency in resource allocation through effects of comparative advantages; and increased innovation, new business processes and new products generated by the dynamics of the internal market.
>
> (Padoa-Schioppa 1987).

In addition, even before complete harmonisation in the EU, several positive effects such as economies of scale are benefits of a more integrated market. The industries which enjoy economies of scale include automobiles and auto component manufacturers, computer firms, office equipment and industrial instrument manufacturers, and the makers of various kinds of non-electrical machinery and chemicals (including pharmaceuticals) (Dunning 1993: 492).

On the other hand, the EU member governments are also concerned in case the SEM leads to a variety of adverse sectoral and regional consequences. For example, they worry about the weakening of indigenous sectoral strengths and diversity, and an increase of external control as a result of the effect of greater foreign competition. But other EU member governments have been equally concerned that 'agglomeration economies may lead to a growing centralization of investment and decision-making in the core regions to the disadvantage of peripheral countries and regions' (Bachtler and Clement 1990: 9). Begg (1989), Clement (1988) and Roegge (1988) pointed out that the SEM proposals raised two problems in the process of integration following EU enlargement. According to them, the SEM proposals raised concern among the EU member states about (1) employees in 'endangered' firms and sectors, as well as (2) regions and member states as a result of changing competition advantages and disadvantages, and an unequal distribution of integration advantages between North and South.

As Thomsen and Nicolaides (1990: 8) suggest, considering all the points made by these writers, the SEM contained some elements which were similar to those of customs unions and hence which may have been expected to have an equivalent effect on trade and FDI. In fact, as far as trade is concerned, the SEM is an example of a customs union. A customs union is an agreement by which members remove all barriers to trade among themselves and adopt a common set of external barriers, which eliminates the need for customs inspection at internal borders (Lindert and Kindleberger 1982: 177).

As Viner explained in his seminal work in the early post-war period, the establishment of a customs union has three main effects on trade: trade creation, trade diversion and trade expansion (or terms of trade) (Viner 1950: 44–5). Trade creation occurs because the elimination of tariffs between the member countries, which decreases costs of firms within the EU, expands their bilateral trade. In other words, lowering costs of firms within the EU will bring the shift from purchasing domestic output to importing from fellow member states which make same products at lower costs. As Thomsen and Nicolaides point out, trade creation would raise social welfare of the importing country:

> Trade diversion occurs when imports from member countries displace more efficient imports from third countries because they are accorded preferential access to the importing country's market. Trade diversion lowers social welfare in importing countries. In fact, the magnitude of trade diversion resulting from the SEM is likely to be small because the main impact of the SEM will be on those service sectors which are not extensively traded internationally.
>
> (Thomsen and Nicolaides 1990: 8)

In other words, service sectors with very little integration across country borders may have much to gain from the SEM, but may not see these gains immediately (Page 1992: 20).

Mansfield (1993: 212) explained the magnitude of trade diversion in terms of political relations. He insisted that trade diversion is likely to be politically beneficial for the alliance partners because one purpose of an alliance may be to divert trade away from neutral or enemy states whose efficiency and political power has been enhanced by trade. He added (ibid.: 211) that 'customs unions formed among allies are likely to enhance the reliability and predictability of trade relations'. According to Mansfield, the EU member states will strengthen the reliability and predictability of intra-EU trade because the EU member states can be regarded as an alliance. Johnson (1965) and Cooper and Massell (1965a, 1965b) explained that customs unions are formed in order to shift imports from outside the union to production within the union for security reasons.

Trade expansion results from gains in efficiency (lower unit cost of production) when firms are able to reap economies of scale by increasing their output. In more detail, the SEM should make the EU a more efficient market, initially through lower costs of transport and easier access to lower-cost goods in other member countries. This in turn should lower EU export prices. Lower export prices enhance EU competitiveness in third markets, all other things being equal (Page 1992: 22). Owing to enhanced competitiveness in third markets, the EU member states could increase their exports, which is trade expansion. 'The welfare effect of trade expansion is more complicated because the gain in efficiency may not be sufficient to outweigh the negative effect of trade diversion' (Thomsen and Nicolaides 1990: 8). However, they are not always comparable. As far as FDI is concerned, a gradual adaptation of 293 measures of the SEM (Dunning 1993: 492) may encourage FDI from non-EU firms for several reasons. Thomsen and Nicolaides (1990: 10) and Han (1992) suggest the following reasons. First, 'the adoption of a common external tariff, through tariff-averaging, may make exporting to the customs-union countries more costly for the foreign firms'.[6] Second, 'even if the level of external barriers remains unchanged, the competitiveness of the foreign firms is reduced relative to firms within the customs union when exporting to a member country. This implies again that exporting becomes more costly'. Third, the EU will emerge as the world's largest single market in the 1990s with a combined population of 322 million[7] and GNP of US$ 3151 billion[8] (Han 1992: 21). 'The creation of a large internal market may enable the operation of large production units benefiting from economies of scale' (Thomsen and Nicolaides 1990: 10) which may then take advantage of increasing purchasing power of the enlarged market (Han 1992: 22). Purchasing power can be increased because the EU's income can grow due to faster EU productivity growth over time and to the one-time improvement in the efficiency of resource allocation in Europe (Oman 1993: 75). Considering these factors, the birth of a large market itself is certainly one element in attracting large investment inflows from outside the EU, although the Chinese market has emerged as a more attractive market for foreign investors.

Fourth, the increased competition among the multinational corporations (MNCs) in the EU is one of the reasons for the expected rise in investment activity. The creation of a single market without border controls among member countries entails the enlargement of the scope of EU competition from a national to a regional level. This, in turn, implies that a company's marketing and production strategies must be extended to cover the whole Union area and that one's home market will be exposed to foreign competitors. In such an atmosphere of intensifying competition, companies will be forced to seek competitive advantage through technical development both to lower input costs and to develop more innovative products. To do this, companies may feel inclined to invest more in the EU.

Fifth, the role of the countries that became members of the EU recently is another reason for the expected rise in investment activity in the EU. Because such EU countries as Spain and Portugal offer relatively lower labour costs, non-EU countries with similar features, like the Asian NICs, could be adversely affected in attracting FDI.

Last, the EU's protectionist trade and trade-related investment policies[9] may have actually 'reinforced the already robust pull-effect of EEI by accelerating the MNCs' strategic diversion of FDI flows from other regions to the EU' (Han 1992: 22).

These six factors could each help to explain a rise in direct investment by firms from non-EU member states. These firms are engaged in FDI in the EU with the main aim of ensuring market access.

As a consequence of the influence of EU integration on imports from developing countries, the firms involved expect that local production (FDI) will eventually increase. In other words, they anticipate that exports from non-EU member countries to the EU will eventually be replaced by local production of these firms (FDI).

5.3 The general perception by non-EU countries of EU integration

There is a big difference in the perception of EU integration between EU member states and non-EU member states. EU member states said that generally EU integration would have positive effects rather than negative ones for non-EU member states.

As briefly explained in Chapter 1 (and later in Chapter 7), however, most bureaucrats and corporate managers in Korea and Japan generally perceive the birth of the SEM as a policy strengthening protectionism because of the EU's measures against these two countries. As Pelkmans (1993: 115) pointed out, in fact, the EU as a whole maintains voluntary export restraints (VERs) with Korea, Japan and Taiwan. In addition, France, Italy, Portugal and Spain also maintain safeguard clauses that sometimes require Article 115 authorisations or VERs if they want to apply the clauses. The quotas (with occasional use of Article 115 authorisations) and VERs of France, Italy and Spain are all directed against Japan and the East Asian NICs. The EU as a whole also employed some common VERs *vis-à-vis* Korea and Japan.[10] Concerning the perception of the SEM by non-EU member states, for example, *Newsweek* reported as follows:

> During the 1980s, when the European Community's Single Market was taking shape and plans were being laid for a common European currency by the year 2000, both prospects were perceived by the Americans and the Japanese as a threat.
>
> (*Newsweek* 12 July 1993: 14)

In more detail, non-EU member states were generally agreed that they got negative rather than positive effects from EU integration. People in non-EU member states also perceived potential threats of EU integration of different types and degrees depending on their situations.[11] Asian NICs, for example, think that they are most threatened from the birth of the SEM. Oman (1993) explained this very well, even though his explanations made numerous assumptions. According to him, the SEM is perceived as the greatest threat by Asian NICs for the following reasons.

First, the exports of Asian NICs to the EU are almost exclusively manufactured goods which have manufacturers in the EU as their competitors, and their trade surplus with the EU also limits the benefits from the terms-of-trade effects[12] at national level (ibid.: 75; Page 1992: 32). In fact, each government has to consider its trade balance, while private firms are concerned about profits in considering the trade relations with foreign countries. Thus, the EU member states tried to maintain the trade balance with Asian NICs as far as possible, which limit the benefits to third countries from terms-of-trade effects.

Second, the SEM preference is for imports of primary products. This may go against many developing countries' interests and against NICs' strategies for increasing their manufactured exports. As of October, 1991, 81 developing countries in Africa, the Caribbean, the Pacific (ACP), and Mediterranean, have co-operation or association agreements with the EU, providing for duty-free access for industrial goods and preferential access for some agricultural products. In addition, the EU has so far provided these countries with financial and technical development aid under the Lomé Conventions of 1975, 1980, 1985 and 1989. The major export products of the ACP countries to the EU are primary products such as tropical crops, sugar, coffee and tea. However, the EU strictly controls the imports from Asian NICs, which are mostly manufactured products, through various measures such as anti-dumping duties, VERs, tariffs and quotas. Oman (1993: 76) added that Latin America and the 'second-tier NICs' of Southeast Asia also could become victims of the SEM as they try to reduce their dependence on exports of manufactured goods to the USA by increasing exports to Europe.

Third, the danger of increased EU use of non-tariff barriers targeted on specific products – notably VERs and anti-dumping duties – is particularly worrisome to the Asian NICs because their exports to the EU increased at the remarkable rate of over 18 per cent during the 1970s and 1980s while the world average increasing rate was below 4 per cent (ibid.: 78). In fact, historically VERs have followed a sudden increase in imports. This is particularly worrisome to Korean firms, because, more than any other Asian NICs, they have been the victims of EU's non-tariff barriers. Korean firms have been weak targets of the EU because neither the Korean government nor the firms prepared for the results of EU policies as much as the other three Asian NICs did. The number of lobbyists working for Korean firms in Brussels is a good example: at the end of 1994, there were less than 10 full-time lobbyists employed by Korean firms.

Fourth, the internal effects of the SEM, or of European integration as a whole, will make EU trade policy more protectionist. Oman has explained why. According to him, particularly significant in this regard is the likelihood that, over time,

> completion of the SEM will induce more re-deployment of industrial activity within Europe from higher-wage to lower-wage production sites, notably to the poorer EU countries and to the ex-socialist countries of Eastern Europe (including East Germany).
>
> (Oman 1993: 77)

In the process of redeployment of industrial activity within Europe, firms in higher-wage EU member countries will suffer from lack of capital and individuals in those countries also may lose their jobs. Therefore, people in the higher-wage countries will put more political pressure on their governments to adopt more protectionist EU trade policies. It is likely that a more protectionist EU trade policy could well be targeted against Asian NICs and other developing countries because of 'ill-founded arguments that EU jobs are being siphoned-off to those countries' (ibid.). As said previously, however, this argument is based on assumptions that the SEM will be completed and it will bring the anticipated results. Thus, it may be that Asian NICs worry unnecessarily. However, Asian NICs do perceive a threat, even if it is only a potential threat. Interviews with managers of Korean consumer electronics firms in the EU certainly gave the impression that this was a strong and commonly held view.

Fifth, the harmonisation under the SEM of national EU import restrictions *vis-à-vis* third countries – mainly the replacement of national quotas by EU-wide quotas for specific products – will have negative effects for non-EU member countries. This is because if the EU member countries agree, regional quotas will increase their effective level of protection. In fact, not all EU countries impose quotas for goods imported from non-EU member countries. The EU imposed definite anti-dumping duties on small-screen CTVs on 17 February 1988. Before that date, Italy did not impose any restrictions on Korean consumer electronics products, while France imposed a quota on Korean CTVs in 1977. The products mainly involved are products of NICs such as footwear, electronic goods, garments and textiles, steel and some tropical products.

Sixth, even investment in local assembly in EU markets does not ensure full competitive access to the single market. Firms from non-EU member countries, in particular lower-income non-EU member countries, cannot benefit from the competitive advantages to be derived from the investment in the EU because of trade-related investment measures (TRIMs). In fact, the EU's TRIMs limit the competitive advantages of those countries that are more likely than their American and Japanese competitors to rely on parts and components produced with low-cost labour at home or within the EU for assembly and sale in Europe. A good example is consumer electronics firms. As of the end of 1991, there were 188 Japanese consumer electronics firms in the EU. Of these, 73 firms (39 per cent) were part makers of electronic and machinery products (JETRO 1992). According to the results of the survey conducted for this research, only four out of 16 Korean consumer electronics firms (25 per cent) in the EU that responded produced parts of electronics products, as of the end of 1994.

Last, the growing *de facto* importance of physical proximity between firms and their customers and suppliers appears more of a worrisome pattern, on the whole, for current or aspiring NICs and developing-country exporters than for their US and Japanese competitors in EU markets. Unlike American multinationals, which have operated in Europe since the 1930s and 1950s, and many Japanese manufacturers which have also moved in the 1980s to set up complete production networks within Europe, the bulk of manufacturers in developing countries and the NICs must rely on low-wage cost advantages at home to compete successfully in European and other developed-country markets through exports.

5.4 The perception by non-EU countries of EU policy towards the consumer electronics industry

EU policy towards the consumer electronics industry, which is strongly influenced by European consumer electronics firms, is a main factor which has forced the location of some Korean and a large number of Japanese plants in EU member states.

Since the 1960s Japanese consumer electronics firms have successfully strengthened their position in the European consumer electronics market. Because of the steady stream of technological innovations (such as the diffusion of digital technology), the development of new types of media (such as direct satellite broadcasting, cable networks, audio and video laser discs, etc.) and the boom in the number of new television stations, this sector has been the scene of fierce competition between European firms and those from Japan and Asian NICs (Commission of the European Communities 1991c: 38 in chapter 12). Since the mid-1980s, consumer electronics firms from Asian NICs have increased their exports to Europe, and they have emerged as major exporters to the EU market since the late 1980s, as Table 5.3 shows. The competitive pressure from Japan and Asian NICs has forced European consumer electronics firms increasingly to concentrate their number by mergers and acquistions (Commission of the European Communities 1990a: 31 in chapter 12). As a result of this pressure, there are three leading consumer electronics firms in the EU member states – Philips-Grundig,[13] Thomson[14] and Nokia.[15]

The responses from local competitors were diverse and ranged from simply facing the competition from non-EU member states' firms to seeking to match Japanese best practice. However, the dominant three European con-

Table 5.3 Consumer electronics structure of imports (1988–92) (ECU millions)

Country	*Imports to EU*				
	1988	*1989*	*1990*	*1991*	*1992*
Japan	5,513 (49.5%)	5,374 (49.0%)	5,334 (44.9%)	5,960 (46.8%)	5,255 (45.1%)
China	353 (3.2%)	591 (5.4%)	850 (7.2%)	870 (6.8%)	955 (8.2%)
Korea	1,228 (11.0%)	936 (8.5%)	1,006 (8.5%)	1,046 (8.2%)	800 (6.9%)
Austria	628 (5.6%)	762 (7.0%)	948 (8.0%)	867 (6.8%)	761 (6.5%)
Singapore	579 (5.2%)	709 (6.5%)	882 (7.4%)	853 (6.7%)	745 (6.4%)
USA	709 (6.4%)	622 (5.7%)	709 (6.0%)	707 (5.5%)	735 (6.3%)
Malaysia	260 (2.3%)	416 (3.8%)	519 (4.4%)	690 (5.4%)	727 (6.2%)
Taiwan	603 (5.4%)	543 (5.0%)	452 (3.8%)	422 (3.3%)	354 (3.0%)
Thailand	—	103 (0.9%)	176 (1.5%)	251 (2.0%)	338 (2.9%)
Hong Kong	456 (4.1%)	298 (2.7%)	266 (2.2%)	307 (2.4%)	227 (2.0%)
Turkey	—	94 (0.2%)	180 (1.5%)	202 (1.6%)	186 (1.6%)
Indonesia	—	—	—	—	88 (0.8%)
Brazil	—	104 (0.9%)	77 (0.6%)	98 (0.8%)	72 (0.6%)
Total extra-EU,	11,133 (100%)	10,958 (100%)	11,881 (100%)	12,743 (100%)	11,643 (100%)
of which EFTA	1,032	1,006	1,229	1,111	941

Sources: *Panorama of EU Industry 1990* (Commission 1990a: 12–31, Table 12); *Panorama of EU Industry 1991–1992* (Commission 1991c: 12–43, Table 8); *Panorama of EU Industry – Statistical Supplement 1992* (Commission 1992a: 12–19, Table 8); *Panorama of EU Industry 1993* (Commission 1993: 10–32, Table 5); *Panorama of EU Industry 1994* (Commission 1994: 10–30, Table 5).

Note: Figures in 1991 and 1992 excluding imports to EU of music recording products.

sumer electronics firms in the EU have adopted two main actions in response to the export offensive of the Far East producers (Cawson *et al.* 1990: 335).

First, European firms have pushed for protection in response to the export offensive of Japan since the 1960s and the Far East producers such as firms from Korea, Singapore, China, Taiwan and Hong Kong since the 1980s. Almost all European consumer electronics firms 'have lobbied vociferously for trade policies which are intended to blunt the competitive edge of the Japanese in combination with other firms, and through industry associations' (ibid.: 218). Since the 1970s, European consumer electronics firms have also lobbied for trade policies which are intended to dull the competitive edge of firms from Asian NICs. As an example, Philips led a campaign by European producers in the early 1980s to persuade the EU to clamp down on so-called 'screwdriver' plants by extending anti-dumping measures from finished products to component imports and by adopting new rules for EU value-added for VCRs (ibid.: 323).

A major change in the strategies of these European firms was the shift in trade policy determinance from national governments to the EU as a whole (Cawson *et al.* 1990: 218). In the 1960s each national government determined these policies. However, since the 1970s, the European Union has made final decision on such policies.

The strategies of the three dominant European consumer electronics firms in the EU were contradictory (ibid.: 333). Publicly, these leading European firms continued to plead unfair practices by the Japanese and Korean consumer electronics firms, while at the same time they sought to establish barriers in Europe to protect themselves against competition. Despite these contradictory strategies of European consumer electronics firms,[16] the EU considered the demands of the European firms. In fact, the Commission shifted its preferred policy instruments from voluntary restraint, through increased tariffs, to action on local content and a stricter interpretation and enforcement of anti-dumping laws. In other words, the demands from European consumer electronics firms created a pressure movement in Brussels towards increased protectionism.

Second, many European consumer electronics firms have adjusted by trying to match Japanese best practice, and by attempting to develop new products and processes. In consumer electronics such strategies have often led to new forms of collaboration in addition to the processes of M&A (ibid.: 335)

In other words, strategies of consumer electronics firms, as in telecoms, have increasingly featured alliances and joint ventures. These strategies are influenced indirectly by the EU. The EU actively promoted the alliances and joint ventures between European firms by launching programmes such as European Strategic Programme for Research and Development in Information Technology (ESPRIT) and Research and Development Programme in Advanced Communications Techniques in Europe (RACE), although these programmes were not directly related to consumer electronics industry in the EU. Government–industry relations have been dominated by links between major firms, national governments and the EU largely as 'cover' for the self-interest of the dominant European producers, Philips, Thomson and Nokia. Trade associations played a relatively minor role (ibid.: 219).

Along with leading European consumer electronics firms, Japanese firms, one of the four main firm clusters in the EU industry[17] have also lobbied for protection in response to the export offensive of the Far East producers, as Table 5.3 illustrates.

Japanese firms facing fierce competition from Asian NICs, in particular Korean consumer electronics firms, were forced to retreat from the lower end of the colour television (CTV) and VCR markets and produce ever more innovative products with higher value-added and higher profit margins. Therefore, 'some managers of Japanese companies even expressed a common interest with European consumer electronics firms in protectionist measures against Korean firms' (ibid.: 327).

In summary, the responses from European firms, in particular the three leading firms and Japanese consumer electronics firms within the EU, made it more difficult for Korean consumer electronics firms to increase their market share in Europe, forcing them to invest in local production in the EU member states.

5.5 The impact of the EU integration on the Korean electronics industry

Since 1985 when the 1992 programme (SEM) was proposed, both Korean FDI and exports to the EU, as a whole, have increased dramatically, as Table 5.4 shows. Both exports and FDI which Korean electronics firms have made in the EU have also shown generally increasing trends, although the export volumes in 1989 and 1992 showed some decrease, as illustrated in Table 5.5 and Figure 5.1. One of the main reasons for the decrease of export volume in 1989 and 1992 was the trade-related investment policies, notably anti-dumping duties, of the EU member states.[18] The another reason for the decrease in export volume was the appreciation of the won. In fact, the won has appreciated 20 per cent during the period of trade surplus (1985 to 1989).[19]

A special phenomenon, occurring temporarily, has been that several Korean consumer electronics firms have recorded an increase in their local production as well as exports from their parent firms in Korea to the EU.

If we analyse the potential effects of the EU integration on Korean electronics firms according to the theories which hold that the relations between FDI and trade are substitutive, the increase of exports and FDI of Korean electronics firms in the EU suggest that EU integration, and in particular the start of the SEM, has not affected Korean exports to the EU member states. However, it is difficult to say that this is true in the case of Korean electronics

Table 5.4 Korean FDI and exports to the EU (1985–91) (exports: US$ billion; FDI: cases/ US$ 1,000)

	1985	*1986*	*1987*	*1988*	*1989*	*1990*	*1991*
Exports (1)							
USA	10.8	13.9	18.3	21.4	20.6	19.4	18.6
EU	3.2	4.3	6.6	8.1	7.4	8.8	9.7
Japan	4.5	5.4	8.4	12	13.5	12.6	12.4
Total exports	30.3	34.7	47.3	60.7	62.4	65	71.9
FDI (2)							
USA (3)			37	50	66	80	84
			(167,705)	(92,896)	(168,520)	(342,627)	(395,244)
EU	2	5	7	14	9	24	21
	(10,562)	(43,859)	(13,157)	(80,423)	(17,688)	(78,193)	(44,078)
Japan			5	15	21	15	21
			(1,279)	(6,503)	(10,179)	(10,925)	(14,934)
Total FDI			49	79	96	119	126
			(182,141)	(179,822)	(196,387)	(431,745)	(454,256)

Sources: 1: Ministry of Trade and Industry; *Trade Policy Review, Vol. 2* (GATT 1992: 73 Table 13 (adapted)); 2: *HaeWae TuJa HyunJi BupIn HyunWhang* [Current Status of Korean Affiliates Abroad] (Bank of Korea 1992).

Notes: 1: On a customs-clearance basis. 2: Total investment. 3: There are some differences in statistics between the 1991 edition and the 1992 edition of *Haewae Tuja TongKae Yunbo* [Overseas Direct Investment Yearbook]. Thus, data in 1989, 1990 and 1991 are based on *Haewae Tuja TongKae Yunbo* (Bank of Korea 1992). Data in 1987 and 1988 are based on *Haewae Tuja Tongkae Yunbo* (Bank of Korea 1991).

Table 5.5 Exports of the Korean electronics industry to the EU member states (1987–92, ECU 1,000)

	1987	*1988*	*1989*	*1990*	*1991*	*1992*
EU total	1,935,818	2,417,606	2,296,039	2,444,833	2,695,575	2,367,052
Growth rate		+25%	−5%	+6%	+10%	−12%
(%)	(87/86)	(88/87)	(89/88)	(90/89)	(91/90)	(92/91)

Source: *Statistics of Electronic and Electrical Industries* (various issues) published by EIAK (Electronic Industries Association of Korea).

Note: There were some differences in the statistic figures of the same year between each edition. In these cases, the statistical figures released by the later edition were used.

firms in the EU. The exports of Korean electronics firms to the EU consist of three sectors: industrial electronics products, components and consumer electronics products.

If we analyse the exports of electronics products of each of these three sectors, export volume of Korean consumer electronics products to the EU member states have decreased, while exports of industrial electronics

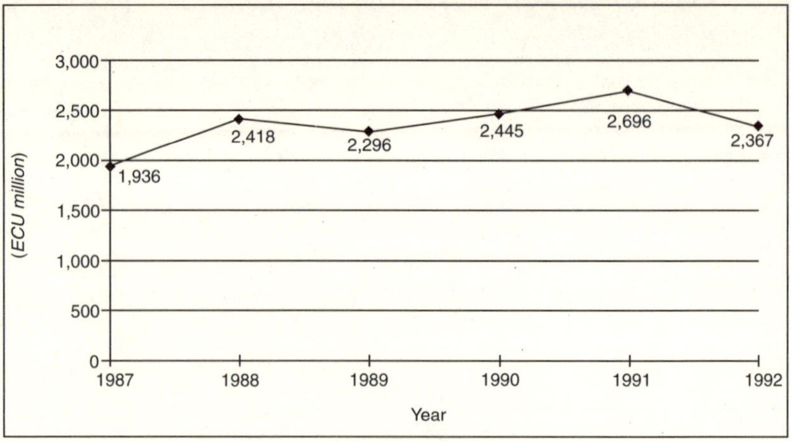

Figure 5.1 Exports of Korean electronics industry to the EU member states.

Table 5.6 Exports of Korean electronics products (industrial electronics products, components and consumer electronics products) to the EU (1988–92, ECU 1,000)

	Industrial electronics products	*Components* (*increase/decrease rate*)	*Consumer electronics products*
1988	711,884	609,933	1,095,789
1989	695,474 (−2%)	710,269 (16%)	890,295 (−19%)
1990	770,532 (11%)	699,561 (−2%)	974,739 (9%)
1991	980,177 (27%)	802,555 (15%)	912,844 (−6%)
1992	1,001,007 (2%)	678,176 (−15%)	687,868 (−25%)

Source: *Statistics of Electronic and Electrical Industries* (various issues) published by EIAK (Electronic Industries Association of Korea).

products to the EU member states have increased, as shown in Table 5.6 and Figure 5.2.

As explained in Chapter 2, Korean FDI in the EU is characterised by a high concentration in the consumer electronics sector. Therefore, the impact of the EU integration on the Korean electronics industries should be analysed by considering the FDI and exports of Korean consumer electronics products. As Table 5.6 shows, exports of Korean consumer electronics products to the EU have decreased, while FDI for those products in the EU has increased.

This can be explained in two ways. First, Korean consumer electronics products were the main victim of non-trade barriers, in particular the anti-dumping duties, of the EU.[20] Therefore, one could expect a reduction in

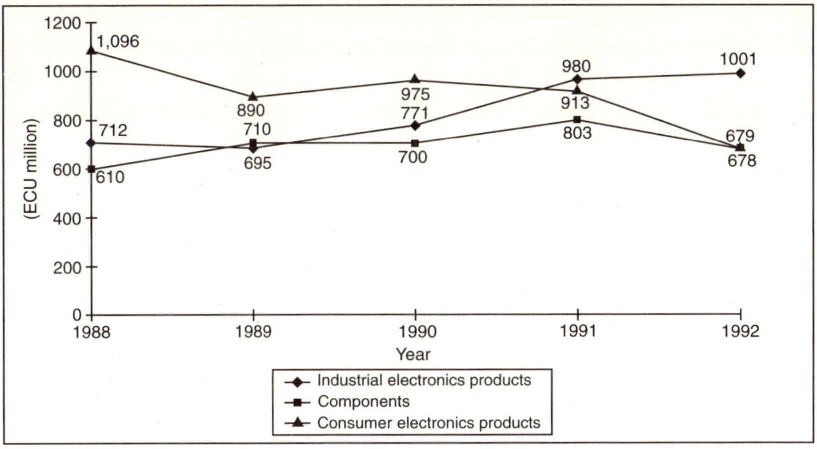

Figure 5.2 Exports of Korean electronics products (industrial electronics products, components and consumer electronics products) to the EU.

the exports of consumer electronics products. Second, Korean consumer electronics firms started to increase their FDI and exports of parts and components in order to avoid trade and non-trade barriers, such as tariffs, quota or anti-dumping duties. These firms, however, do not have complete production networks within Europe, and therefore the imports of components from Korea have increased. As Oman (1993: 79, 81) points out, Korean firms did not invest in the production of industrial electronics products and components in the EU member states because of the short period of FDI in the EU. In other words, the only way open to Korean electronics firms in the EU market is to 'remain heavily dependent on their ability to import components that are mass-produced in lower-wage Korea, in order to gain price competitiveness' (ibid.: 81), because Korean consumer electronics firms do not set up complete production networks within Europe, unlike American or Japanese companies. As will be discussed in Chapter 8, as of the end of 1994, nine of the responding 16 Korean consumer electronics firms[21] (56 per cent) in the EU produced end-user products, while only four companies[22] (25 per cent) produced components. Only one company[23] (6 per cent) produced semi-finished products and the other two companies[24] (12.5 per cent) produced both semi-finished products and end-user products. However, according to the Japan External Trade Organisation (JETRO 1992), 73 out of 188 cases of FDI in electronic and electrical machinery were made by electronic and electrical machinery part manufacturers as of the end of January 1992, such as, for example, Matsushita Electric (UK) Ltd. Matsushita, which manufactures and does R&D for CTVs and microwave ovens, started production in 1974, and Matsushita Electric Components (UK) Ltd, which manufactures and markets electronic components, started production in 1988. Because of the lack of complete production networks within Europe, the export of parts or components has increased in relative terms, while the export of end-user consumer electronics products from Korea has decreased. However, the export of components is also controlled by the EU's local-content requirements or rules of origin.[25] Certainly, in the long run, this is one of the most important reasons why Korean electronics firms should increase local production.

In summary, in line with the general influence of EU integration on FDI, EU integration has already brought and will continue to bring an increase in FDI by Korean electronics firms in EU member countries. Eventually, local production is expected to replace the exports of Korean electronics firms to the EU member states. The results of the survey and the interviews with senior managers of nine Korean electronics firms in the EU confirmed this.[26]

As previously explained, several Korean consumer electronics firms have recorded an increase in their local production as well as exports from their parent firms in Korea to the EU. This can be explained as follows. First, the absolute volume of exports of those firms to EU member states increased due to Korean firms' efforts at diversifying their export markets. Along with the Chinese market, the EU market has certainly emerged as an important one for Korean firms because of the protectionism of certain countries, especially the USA and Japan, and other attractive features of the EU market, such as its sheer size. After the American government and the Japanese government restricted imports from Korean firms in the 1980s, Korean firms made every effort to maintain or increase their exports by diversifying into world markets. Consequently, the absolute export volume of Korean consumer electronics firms in the EU has increased since the mid-1980s, as Table 5.5 shows.

Second, the interviews with senior managers of nine Korean consumer electronics firms in the EU revealed that they started local production in places with the most favourable conditions[27] in order to maximise their business efficiency until such time as they set up production networks in the EU. In effect this meant that they produced some products in the EU if the production conditions in the EU were more favourable than in Korea or other locations.

In summary, the SEM will result in a partial replacement of exports by Korean electronics firms to the EU with local production. For the transition period of shifting from exports to local production, there are several firms that show an increase both in their exports from Korea to the EU and in their production from plants in the EU. However, according to interviews with senior managers of Korean electronics firms in the EU member countries, a good deal of the exports of these firms will eventually be also replaced by local production, because of various EU regulations, such as local-content requirements, rules of origin and so-called 'screwdriver-plant regulations', or simply because they are both competitive and secure inside the EU.

5.6 Conclusion

The start of the SEM brings both positive and negative trade effects for non-EU member countries, even in the absence of any change in EU trade policy (Oman 1993: 75). However, it can be said that governments outside the EU perceive EU integration (especially the birth of SEM) as protectionist, even when the governments of the EU member states try hard to moderate this image. Korean firms, including consumer electronics firms, are no exception. As Klein and Hong (1993) pointed out, because of their experiences in trying to remove barriers to agricultural trade under the Uruguay Round, non-EU member countries are quite pessimistic that the EU member countries will liberalise trade with them and not impose new barriers. The EU has also restricted the imports of textiles and clothes, traditionally one of the major exports of ASEAN countries through the Multi-fibre Agreement (MFA) and its bilateral enforcement by EU member countries. This is another reason why non-EU member countries find it hard to be optimistic about EU integration.

In addition, both the protectionist measures taken by the American government and the attractiveness of the SEM, in particular the sheer size of the market, could make non-EU member countries treat its birth more seriously.

The completion of the SEM will also partially replace exports of Korean electronics firms by local production. In the case of Korean electronics firms, the explanations offered by Mundell (1957) and Rock (1973) are correct. Korean exports of electronics goods to the EU have faced various trade or non-trade barriers. Korean electronics firms started to invest in production in the EU in order to maintain their market share by avoiding trade and non-trade barriers.

Korean FDI in the EU is concentrated in the consumer electronics sector. Therefore, the analysis of the impact of the SEM on Korean electronics firms in the EU can be made by investigating the relations between exports and the number and value of FDI by Korean consumer electronics firms. Considering the relations between the cases of anti-dumping duties against Korean consumer electronics products and their export volume, and cases of FDI by Korean consumer electronics firms in the EU, the rapid expansion of local production of Korean consumer electronics firms in the EU was certainly in response to factors external to the firms – the strengthening protectionism of the EU coming as a result of the SEM, including anti-dumping duties, not internal *to* the firms. In fact, anti-dumping duties against Korean consumer electronics products started to increase in 1986, and the number and value of FDI by Korean consumer electronics firms in EU member states started to increase in 1987–8. In addition, exports of industrial electronics products and components of Korean firms started to increase in 1988, while exports of consumer electronics products of Korean firms started to decrease in 1988.

This was also proved by the results of the survey.[28] The survey results showed that the major motivation for companies was the desire to become multinationals. However, three out of the top five motives were to do with avoiding the protectionist measures taken by the EU: the second motive was to avoid discriminatory restrictions such as tariffs and quotas; the fourth was the fear of strengthening protectionism as a result of the EU's market integration and the fifth was to avoid anti-dumping decisions against Korean-made components. Considering this, it looks as if the rapid expansion of local production by Korean consumer electronics firms in the EU was in response to factors external to the firms – the strengthening protectionism of the EU coming as a result of the SEM including anti-dumping duties.

There are limits to how closely one can analyse relations between FDI by Korean electronics firms in the EU and exports from Korean electronics firms to the EU. These limits are similar to those outlined by Thomsen and Nicolaides (1990) in their study of the relations between Japanese FDI in the EU and Japanese exports to the EU. In both cases there is an absence of a long data series (particularly true in the Korean case), and in addition the lower level of overseas production of Korean consumer electronics firms, severe restrictions against Korean exports in electronic goods sectors and the negligible volume of Korean FDI in the EU before the mid-1980s must all be taken into account. The latter can make the growth rate easily skewed by just one or two firms.

Despite these limitations, based on the data available, it can be concluded that relation between FDI and exports by Korean electronics firms to the EU is substitutive in a limited sense and that the rapid expansion of local production of Korean consumer electronics firms in the EU was mainly a response to factors external to the firms, rather than internal factors such as trade imbalances, wage increases and appreciation of the won. In other words, Korean consumer electronics firms started to invest their local production in the EU not for offensive reasons but for defensive reasons.

6

EU TRADE POLICIES (COMMON COMMERCIAL POLICY)

The strategy of investors is influenced not only by the investment measures of their home countries. The policies of host countries, especially trade policies and policies related to inward FDI, such as investment incentives and competition policy, are also so-called 'pull factors' for foreign firms considering making a foreign direct investment.

Generally, the EU's trade policies have strongly influenced whether or not a foreign firm invests in the EU. By contrast, the investment incentives offered by individual EU member states have had a stronger influence on decisions relating to location, size and timing of investments. At the time of the opening of the SEM in Europe, some changes had occurred or were occurring in the trade policies of the EU member countries. These influenced the investment decisions of firms from non-EU member countries.

This chapter shows that the common commercial policy of the EU can still be characterised by instability, discrimination and lack of transparency (Hiemenz *et al.* 1994: 56; Tsoukalis 1993: 288–90). In an interview on 6 December 1993 Mr Stephen Woolcock, Senior Research Fellow in the European Programme at the Royal Institute of International Affairs agreed that the common commercial policy of the EU was uncertain. The records of the EU's common commercial policy, which is analysed in this chapter also show instability, discrimination and lack of transparency. This is the main reason why Korean consumer electronics firms have started local production within the EU.

The characteristics of the EU's external trade policies will be reviewed in full detail by answering the following questions: 'What changes have there been in the EU's trade policies?' and, 'How have these policies themselves influenced the decisions of firms of non-EU member states, in particular Korean consumer electronics manufacturers, when considering making direct investment in the EU?'

This chapter is organised into five sections covering the EU's common commercial policies at national level and then at Union level; the EU's policy towards the consumer electronics industry; EU policy-making procedures and finally a conclusion.

6.1 The EU's common commercial policy at national level

Any analysis of developments in Union trade policy must take into account the wider international context. Since the late 1980s, there has been a more protectionist approach to trade policy generally in developed countries due to the fact that since the mid-1970s NICs have emerged as strong competitors of advanced countries in the world market.

The EU's trade policies (the common commercial policy) are based on Articles 110–16 of the Treaty of Rome. In the European Union, trade policies are made at two levels: national level and Union level. Because of this, it can be said that there are still no truly common EU external trade policies.

Questions concerning trade policies of EU member states at national level can best be answered by an examination of trade policies in some of the larger countries of the EU. Therefore, in this section trade policies of three larger EU member states (France, Germany and the United Kingdom) are analysed at national level.

France is a representative country that has been at the protectionist end of opinion in the Union. France has advocated protectionist policies by the Union as a whole and has been prepared on occasions to take unilateral action, in defiance at least of the spirit of Union law (Hayes 1993: 88). For instance, in June 1977, the French authorities were in the forefront of those who pushed for tighter restrictions on imports of textiles and clothing from non-EU member countries under the MFA. At that time, the French government notified the EU Commission that it was going to take unilateral action to impose quantitative restrictions on imports of various textiles and clothing items from LDCs.

This aside, the French government showed its protectionist nature in several other cases. In the first case, imports of cars from Japan were limited to 3 per cent of the French market in 1977. Then, by the decree of the Minister of the Budget, in 1982 all imports of VCRs from all sources had to be cleared through the small customs post of Poitiers. The third case is the fact that, in the early 1980s, the French government made great efforts to adopt the Union's New Commercial Policy Instrument which was designed to facilitate action against illicit commercial practices by non-EU member states. The protectionist nature of French trade policy is still stronger than the other larger EU member states. As an example, the French government restricted five categories of Korean goods at of the end of 1991, which was the largest number of restrictions among developed EU member states (Korean Foreign Trade Association April 1992). The five categories are radios, CTVs, silk fabrics (restricted by quota allocated by country), electric and electronic equipment (restricted by global quota), and regenerated long-fibre goods (restricted by bilateral quota).

By contrast, successive German governments have argued consistently for more liberal import policies for manufactured goods within the EU (Porter

1990b: 378). The German position, as it was then and as it has remained essentially the some throughout the 1990s, was set out in a memorandum sent to the Council of the Union in 1978 by Count Otto von Lambsdorf who was German Economics Minister from 1977 to 1984.

It has to be noted that these statements of liberal economic principles apply more to the industrial sector than to some other parts of the German economy, notably agriculture, insurance, banking and trade in services, even within the Union itself. Hayes (1993) pointed out that five elements worked together to support free market attitudes, particularly in relation to manufacturing in Germany. According to him, a conscious decision to adopt a generally free market approach after the Second World War, encouraged by the US occupation, was the first element. The economic miracle of the earlier post-war years gave greater momentum to the approach. Moreover, both the administrative and political machines insulated bureaucrats treating economic issues from political pressure. The broader German institutional structure has also been such as to put supporters of the social market orthodoxy in positions of power and influence (ibid.). Finally, despite a change in the dominant party in the federal government,[1] the necessity of coalition-forming since 1972 led to the appointment of Federal Economics Ministers drawn from the right wing of the minority Free Democratic Party, which has consistently favoured liberal trade policies. For example, as of the end of 1991, the German government restricted the import of only one category of goods (canteens of cutlery) from Korea. This was the smallest number of restrictions among advanced EU member states (Korean Foreign Trade Association April 1992).

Until the early 1980s, the position of the United Kingdom could be said to lie in the middle between the protectionist and liberal wings of the EU member countries (Hayes 1993: 105). That is, until the early 1980s 'the policies of successive governments in the United Kingdom on imports of manufactures tended to be nearer to the French than to the German position' (ibid.: 105). In the early 1990s, this tendency was still the case for the United Kingdom as a whole.

The liberal aspect of British government policies can be seen in its support of a new round of multilateral trade negotiations in the General Agreement on Trade and Tariffs (GATT) in 1985. This was in fact against the wishes of the French government. In the early 1990s, the British Conservative government preferred a generally liberal outcome to the Uruguay Round.

The prevailing orthodoxy in the United Kingdom emphasised the potential gains from unrestricted international trade. Since the end of the Second World War, there have been changes in the approach taken by officials dealing with trade policy. In the early post-war years, they followed a generally liberal approach. In the 1970s, however, there was a move among officials to a more pragmatic approach, in parallel with the post-oil shock recession and the consequent rise of protectionist pressures at the political level. 'In the second half of the 1980s and early 1990s, there have been signs of a more liberal approach to trade policy among officials, again in parallel with the evolution of attitudes among ministers' (ibid.: 106). As an example, at the end of 1991, the British government, like that in Germany, restricted the import of only one category of goods (again, canteens of cutlery) from Korea (Korean Foreign Trade Association April 1992).

6.2 Trade policies at Union level

It is generally accepted that despite national differences, the EU does have a common trade policy *vis-à-vis* third countries. And an essential element of the Treaty of Rome was the common external tariff (Balasubramanyam and Greenaway 1993). The Treaty of Rome, by which the EU was founded, was explicit on matters of trade, while vague and mostly silent on monetary matters.

In Article 110, it is stated:

> By establishing a customs union between themselves the Member States aim to contribute, in the common interest, to the harmonious development of world trade, the progressive abolition of restrictions on international trade and the lowering of customs barriers.[2]

In addition to a common external tariff, certain variations in the level of the tariff, and exceptions to it (Article 115), are negotiated at the Union level. Until recent years, however, this common policy has been eroded by national measures, most notably source-specific restraints such as VERs. According to Article 115 of the Treaty of Rome, special exceptions have been made allowing purely national measures.

In Article 115, it is stated:

> In order to ensure that the execution of measures of commercial policy taken in accordance with this Treaty by any Member State is not obstructed by deflection of trade, or where differences between such measures lead to economic difficulties in one or more of the Member States, the Commission shall recommend the methods for the requisite co-operation between Member States. Failing this, the Commission shall authorize Member States to take the necessary protective measures, the conditions and details of which it shall determine.
>
> In case of urgency during the transitional period, Member States may themselves take the necessary measures and shall notify them to the other Member States and to the Commission, which may decide that the States concerned shall amend or abolish such measures.

> In the selection of such measures, priority shall be given to those which cause the least disturbance to the functioning of the common market and which take into account the need to expedite, as far as possible, the introduction of the common customs tariff.

In fact, Article 115 of the Treaty of Rome is the main legal instrument which is used in order to prevent the free circulation of goods imported from abroad to the Union. There is, however, a safeguard clause under which the Commission may allow a member state to take protection measures against other member states in order to prevent imports from non-EU member states from being diverted elsewhere in the Union. In 1982, the UK Department of Trade and Industry stated the Union's view on Article 115. According to the statement, 'Article 115 may only be used in respect of products subject to restraint by formal quota arrangements; it cannot be used against goods subject to voluntary restraint or other informal arrangements' (*Journal of Common Market Studies* 28(3), March 1990: 206).

However, after 1993 the Commission was determined to stop this. The signing of the Single European Act, in practice, represented the start of a commitment by member states to operate the rules of the treaties in full. Until the signing of the Single European Act[3] in 1986, the member states had applied only some provisions of the Treaty of Rome, while they also decided not to apply the rest, such as articles 74–84 on a common transport policy.

However, even after the signing of the Single European Act, it can be said that the EU member states have not decided whether to implement the whole package at some point. In addition, even if the Treaty is implemented in full, there has always been a tacit understanding that the Treaty of Rome will be applied in an expedient manner if this is deemed to be politically necessary.

Thus, in order to have a truly common trade policy at Union level, the Union was supported in a bid to harmonise a range of policies and to rephrase Article 115 provisions. These measures reduce the scope for independent action on the part of individual member states. In this section, therefore, the trade policies followed by the European Union from its formation in 1957 up to the early 1990s are reviewed in more detail by considering the following topics:

1 anti-dumping and anti-subsidy measures (EEC Regulation no. 2423/88);[4]
2 the Union's external tariff (EEC Regulation no. 2886/89),[5] together with various preferential arrangements for countries outside the Union;
3 quantitative restrictions and restraint agreements with supplier countries;
4 the New Commercial Policy Instrument adopted in 1984 (EEC Regulation no. 2641/84)[6] and the common rules for imports (EEC Regulation no. 288/82).[7]

6.2.1 *The anti-dumping policy and the anti-subsidy policy of the EU*

The rules of the Treaty of Rome already ban predatory pricing, in other words, artificial price cutting in order to drive a competitor out of business and preserve or create a monopoly position. And GATT, which is the body of rules governing world trade, defines dumping as a situation where:

> The price of a product exported from one country to another (a) is less than the comparable price, in the ordinary course of trade, for the like product when destined for consumption in the exporting country, or (b) in the absence of such a domestic price is less than either (i) the highest comparable price for the like product for export to any third country in the ordinary course of trade, or (ii) the cost of production in the country of origin plus a reasonable addition for selling costs and profits.
>
> (GATT 1947, Article VI)

The European Union first made common anti-dumping regulations which were against dumped or subsidised imports from non-EU member states, in 1968, although they were not used until 1976.

The regulations were set out in Council Regulations (EEC) No. 459/68[8] in accordance with existing international obligations, in particular those coming from Article VI of the GATT and from the first Agreement on Implementation of Article VI which is the 1968 GATT anti-dumping code.

In fact, Article VI of GATT is more widely used by the EU and the USA than Article XIX of GATT, which allows countries to impose 'safeguard' duties when 'unforeseen developments . . . cause or threaten serious injury to domestic producers'.[9] This is because Article XIX of GATT imposes strict conditions on 'safeguard' measures and obliges the country taking action to justify itself. In addition, Article XIX obliges an importer to impose duties without discriminating between suppliers, while Article VI allows importing countries to impose duties country by country, or even firm by firm, if it can be shown that imports are being 'dumped', and if it can be shown that domestic producers have been 'injured' as a result (Holmes 1991: 16). In other words, the uncertainty of the EU's common commercial policy was magnified by these ambiguous regulations.

The regulations have been revised on a number of occasions since 1968, as Table 6.1 shows.[10] Current policy is based upon Council Regulation no. 2423/88 of 11 July 1988. The regulations are written in accordance with the GATT Charter and the 1979 GATT anti-dumping code.

According to the regulations, first the European Commission is required to act on any complaint of dumping brought by Union producers of the product concerned, when they consider themselves to have been injured or threatened by dumped imports. The Commission must then determine whether there is

Table 6.1 Amendments of Council Regulations (EEC)

Council Regulation	*Year*	*Aim*
EEC No. 459/68	1968	• In accordance with existing international obligations, in particular those arising from Article VI of the GATT and for the first Agreement on Implementation of Article VI (1968 Anti-dumping Code).
EEC No. 2011/73	1973	• To rationalise certain procedures.
EEC No. 1411/73	1973	• To allow the Commission, where necessary, to alter or withdraw anti-dumping measures or subsidies instituted on a national basis by the UK or Ireland during the transitional period laid down by the Treaty of Accession.
EEC No. 1681/79	1979	• To define more clearly the rules for determining the injury sustained by a Union industry affected by dumping, and the conditions under which interested parties might have access to information used during investigations. To provide that provisional duties might be definitively collected, even when exporters had given voluntary undertakings to revise their export prices. • To provide for the EU Commission to undertake dumping investigations if so requested by a Union industry which considered itself injured or threatened by dumped or subsidised imports.
EEC No. 3017/79 EEC No. 1580/82	1979 Jan. 1980	• Article 14 of the Regulation was amended so that reviews of Union measures could only be requested if at least one year had elapsed since the conclusion of the previous investigation.
EEC No. 1761/87	Jun. 1987	• To extend Article 13 so that an anti-dumping duty could be applied not only to imported finished products, but also to imported components.
EEC No. 2423/88	July 1988	• To clarify procedures, particularly the calculation of normal value and export prices, and Article 13 to prevent exporters from nullifying the effects of anti-dumping duties by bearing the costs themselves.

sufficient evidence to justify initiating an investigation. In other words, an anti-dumping investigation must first establish the existence of dumping.

Second, during the investigation, the Commission may impose provisional anti-dumping duties for a period that does not exceed four months. If requested, the period for provisional anti-dumping duties may be extended for an extra two months. If the Council of Ministers decides against definitive duties, provisional duties, which take the form of a security paid by the exporters in question, must be released after four months or six months if an extension has been granted. An investigation into a dumping case, however, may be dismissed if the complaint is not supported by evidence.

Alternatively, the Commission may accept a price undertaking offered by the exporters. The undertaking may involve a revision of prices so as to eliminate the margin of dumping or an export restraint. One of the characteristics of EU anti-dumping policy is that many cases have been concluded with price undertakings of the exporters concerned (Montagnon 1990: 80).

If the investigation produced evidence of dumping and of injury, or threat of injury, to Union producers and no undertakings are offered or accepted, the Council of Ministers can impose definitive duties on a proposal submitted by the Commission. The rate of duty must not be more than the margin of dumping that has been established, but it may be less. Both anti-dumping duties and exporters' undertakings are subject to review at any time, either on the initiative of the Commission or at the request of a member state. In summary, an anti-dumping investigation must first establish the existence of dumping and provide an estimate of the margin of dumping. The anti-dumping actions taken by the EU have provided non-EU member states with strong reasons for worrying about the emergence of a 'Fortress Europe', due to the problems in the procedures regarding the EU's anti-dumping duties. In fact, the anti-dumping duties of the EU have caused much controversy in recent years, although the purpose of anti-dumping actions, at least, is to help ensure fair competition in the Union. Many economists, such as Hindley (1988) and Messerlin (1990), have argued that anti-dumping policies are the cause of more unfairness than that which they seek to eliminate (Holmes 1991: 17).

The anti-dumping policies have made non-EU firms think that the start of the SEM strengthens the image of Fortress Europe for the following reasons. First, there is a problem in the EU's calculation of the margins of dumping. The EU does not look at the domestic and export prices actually charged by firms. The margin of dumping is the difference between the export price and the domestic price. The approach adopted in the EU is to express both the export price and domestic price at the ex-factory level. To get the correct margin of dumping, the export price and domestic price must be compared at an equivalent level of trade. As Hindley (1988: 449–50) argued, however, the export price and the domestic price are being compared at different levels of trade. This calls for explanation.

To obtain the ex-factory price, all costs incurred between importation and resale, plus a reasonable margin for profit should be deducted. All costs include all the distribution expenses of the sales subsidiary in the importing country such as rent, salaries, advertising, R&D costs, etc.

The export price is calculated by taking the price at which the imported product is first resold to an independent buyer. The selling price should be used to estimate the export price only when the sales company sells the product to another dealer or distributor not related to the exporting company, or directly to the consumer.

The domestic selling price must be adjusted for any import charges or indirect taxes paid and selling expenses. The domestic selling price includes all transport, insurance, handling, loading and ancillary costs, costs of packing the product, the costs of any credit granted, the costs of providing quarantines, guarantees, technical assistance and other after-sales service and other selling expenses, such as any commissions paid or salaries of sales people.

However, when calculating the ex-factory domestic price, the Union may deduct only some selling expenses, which is a problem. Selling expenses which should be deducted, should include advertising or research and development costs and the profit margins of the dealer or distributors where these are related in ownership to the product. But the Union does not do this. As a result, the investigation may tend to find evidence of dumping when in fact no dumping is taking place.

Second, there is a problem with the method used in assessing the export price. Very often a product is sold in the importing country at a range of different prices. Then, the problem lies in the fact that the EU ignores all sales above the home market reference price, instead of taking the average of these prices. What the EU Commission does is to obtain an average price only by taking the average of export prices below the reference price. Once again this increases the possibility of proving that dumping is taking place when in fact it is not (Hindley 1988: 449–54; Norall 1986: 101–5).

Third, if there is no comparable domestic price, the procedure to get the normal value is excessively complex so that the possibility of error increases (Hindley 1988: 451–2). The number of calculations involved and the volume of information required is especially great when no suitable domestic price exists and a normal value has to be constructed. This requires adding to the costs of production a reasonable margin for profits. The normal value is 'the comparable price actually paid or payable in the ordinary course of trade for the like product intended for consumption in the exporting country or country of origin'.[11] The phrase 'in the ordinary course of trade' means that the normal value is determined by taking the first sale of the product to an independent buyer. In other words, if the producer sells the product to a dealer or distributor with whom it is associated, the transaction is not valid for the purposes of determining the normal value. It is only valid to use the selling price to estimate normal value when the dealer or distributor sells the products to an independent buyer.

If there were no or insignificant sales of the product on the domestic market of the exporting country, the producer takes either the price of the product when exported to some third country or to construct a value by adding the costs of production and a reasonable margin for profit.

If the exporter does not produce or sell the product at all in the country of origin, a value is constructed based on the prices or costs of other sellers or producers in the country of origin.

Then, the prices or costs of other sellers or producers in the country of origin becomes a purely hypothetical figure, which can be very different depending on the assumptions made. If the prices are fixed too high, the result will be an inflated normal value for the product.

Fourth, the existence of injury or the threat of injury to Union producers is alleged but is hard to prove with any certainty or objectivity. Once dumping has been established, the Union should prove the existence of injury or the threat of injury to Union producers. It is very difficult to demonstrate that the dumped imports have caused the injury because, very often, there are a number of different reasons for the difficulties being experienced by an industry. This means that whether an industry is facing an injury or the threat of injury entirely because of imported goods or not might be decided subjectively. One representative example of this is the case of textiles in the 1950s.

The Commission tries to prove the existence of injury or the threat of injury to Union producers by considering a variety of factors, such as the volume of dumped imports, their rate of increase relative to domestic production or consumption and the effects on the domestic industry such as production, capacity utilisation, stocks, sales, market share, price, profits, return on investment, cash flow and employment. What it is complaining of very often is the rate of market 'invasion' not its 'fairness'.

Fifth, the so-called 'screwdriver-plant regulation' is a problem in itself because of the absence of any universally agreed definition. The regulation is one of the most controversial in EU anti-dumping regulations. It came into being in June 1987 under Council Regulation 1761/87.[12] The term 'screwdriver plant' has come to be used to refer to plants set up by foreign producers inside a Union member state to assemble finished products using components and parts imported from the home country. There is then, no clear, universally agreed definition of a 'screwdriver plant' except that the value-added locally to the product is low.

The problem comes from the Commission's interpretation of these provisions. The Commission assumes that any local content, i.e., value-added locally, of less than 40 per cent of their total value, constitutes dumping. Despite these problems, defenders of European anti-dumping policy often claim that those who feel mistreated by the process can find a solution for problems by taking the matter up with the European Court of Justice. They also insist that the Court has so far found no fault with it. However, the Court does not make qualitative judgments about policy itself. The

European Court of Justice has simply reviewed whether the correct procedures have been followed.

In conclusion, the use of anti-dumping measures can discriminate in practice against foreigners, making it more difficult for them than it is for indigenous firms to gain access to the market. In such circumstances, the imposition of anti-dumping duties is likely to impede something that normal competition policy would allow. In short, as Montagon (1990: 78–9) emphasises, while anti-dumping measures may be justified as preventing predatory pricing, they actually go much further.

(i) Anti-dumping duties of the EU on Korean products

There are several characteristics of the anti-dumping duties of the EU on Korean products. First, the EU imposed a number of anti-dumping duties on Korean products. As Table 6.2 shows, as of the end of 1991, the EU imposed anti-dumping duties on 14 Korean products, which was the third-largest number of cases (Commission of the European Communities 1991a).

Second, the EU imposed anti-dumping duties primarily on Korean consumer electronic products. As Table 6.3 shows, from 1987 to November 1992, the EU opened 21 anti-dumping investigations. Out of 21 cases, the EU imposed provisional or definitive duties in 10 cases. Depending on the dumping margin calculation made by the EU, definitive dumping rates are decided so that the difference of rates imposed on each firm are big. In four cases, it concluded with an undertaking to raise prices. In four cases the investigations had been dismissed for lack of evidence as of November 1992. The EU was still investigating three cases. Interestingly, in six cases of 10, where the EU imposed definitive anti-dumping duties, and one case of four, where the EU concluded with a price undertaking, the restrictions referred to Korean consumer electronics products.

Table 6.2 Anti-dumping duties imposed by the EU (1987–91)

	1987	*1988*	*1989*	*1990*	*1991*	*Total*
Japan	7	4	2	3	5	21
China	—	7	5	4	4	20
Korea	5	4	1	3	1	14
Taiwan	3	1	1	1	2	8
Hong Kong	3	3	2	—	—	6
Thailand	—	3	—	2	—	5

Source: *Annual Report of the Commission on the Community's Anti-dumping and Anti-subsidy Activities* (Commission of the European Communities 1991a).
Note: Classify according to the investigation opening date.

Table 6.3 Anti-dumping duties on Korean products (1987–92)

Product	*Investigation opening date*	*Results*
Oxalic Acid	22 May 1987	Definitive duties imposed (20.21%, 15 July 1988)
Partially Textured Polyester Yarn (PTY) and Partially Oriented Polyester Yarn (POY)	1 July 1987	Provisional duties imposed (17 June 1988) Definitive duties imposed (PTY 8.1%, POY 0%, 16 Dec. 1988)
CDP	7 July 1987	Provisional duties imposed (18 July 1988) Definitive duties imposed (10.7–26.1%, 17 Jan. 1990)
VCR	27 Sep. 1987	Provisional duties imposed (31 Aug. 1988) Price undertaking (28 Feb. 1988)
Video tape	18 Dec. 1987	Provisional duties imposed (24 Dec. 1988) Definitive duties imposed (22 June 1989, 1.9–3.8%)
Inner tubes and new tyre cases for bicycles, sewn-in inner tubes for racing bicycles	May 1988	Price undertaking
Small-screen CTVs[a]	17 Feb. 1988	Provisional duties imposed (28 Oct. 1989) Definitive duties imposed (27 Apr. 1990, 10.2–19.6%)[b]
Glutamic acid and its Salts	4 June 1988	Price undertaking (3 Mar. 1990)
Album	15 Dec. 1988	Price undertaking (31 May 1990)
Audio cassette tapes	14 Jan. 1989	Provisional duties imposed (Oct. 1990) Definitive duties imposed (14 May 1991, SunKyung Ind. 2.6%, GoldStar 9.2%)
Non-refillable pocket gas lighters	7 Apr. 1990	Definitive duties imposed (25 Nov. 1991, 22.7%)
Car radios	8 May 1990	Definitive duties imposed (Feb. 1992, 6.3–38.3%)
Synthetic Staple fibre of Polyester, not carded, combed or otherwise processed for spinning	2 Nov. 1990	Provisional duties imposed (16 July 1992, SunKyung Ind. 1.6%, SamYang 9.0%)
DRAM (dynamic random access memory)	6 March 1991	Provisional duties imposed (18 Sep. 1992, 10.1%)
Electronic scales	4 Apr. 1992	Under investigation
Floppy disks	18 Sep. 1992	Under investigation
Large-screen Colour TVs[c]	20 Nov. 1992	Under investigation

Source: Updated from Overviews on *Import Restrictions of Major Industrialised Countries* (Korea Foreign Trade Association April 1992).

Notes: a: Small-screen colour TVs include TVs with a screen which is smaller than 16 inches. b: After the re-examination on the rate of definitive duties which started in February 1991, the Commission reduced the rates imposed on Korean manufacturers except for DaeWoo, GoldStar and Samsung from 19.6% to 10.5% in September 1991. c: Large-screen colour TVs include TVs with a screen larger than 16 inches.

Third, anti-dumping duties were the major method that the EU used in order to restrict the imports of Korean products to the EU member countries. According to the Korea Foreign Trade Association (April 1992), as of the end of 1991, the EU restricted the import of 14 Korean product categories. Among these, the EU imposed anti-dumping duties on 10 Korean product categories to restrict imports.

(ii) *Anti-subsidy actions*

'By comparison with anti-dumping, the Community makes very sparing use of countervailing duties against products benefiting from unfair subsidies paid in third countries' (Montagnon 1990: 81). The European Union, in fact, initiated 10 anti-subsidy actions between January 1980 and December 1989 and introduced five measures. According to Montagon, there are three reasons why the Union uses countervailing duties relatively less than anti-dumping actions.

The first is that, unlike their US counterparts, European officials consider it hard to prove that when a subsidy has been granted, it has a demonstrable and direct effect on trade. The second is that subsidies are relatively uncommon in those parts of the world, such as the Far East, which have recently become Europe's most serious cause for concern on trade policy. The third reason, which is probably the most important, is that the Union is vulnerable; it does not want any criticism on Europe's own subsidies, not only in agriculture but also in certain sectors of manufacturing. In other words, factor price differentials make government subsidy unnecessary.

The GATT permits members to levy anti-dumping duties to restore fairness in trade. Similarly countervailing duties may be imposed where it can be shown that a foreign exporter is in receipt of some kind of government subsidy. The European Union's anti-dumping and anti-subsidy provisions are described in regulation no. 2423/88 (NCC 1993: 76).

Commission officials consider anti-subsidy provisions, together with the anti-dumping provisions as one of the most important instruments of its common commercial policy (GATT 1991). In sum, the regulation and its application have been subject to much criticism in recent years due to the various problems outlined above.

6.2.2 *The Union's external tariff, together with various preferential arrangements for countries outside the EU*

Tariffs used to be one of the main instruments of protection. However, their importance as an instrument of protection has drastically declined in all industrial countries as the result of a series of post-war trade negotiations under the GATT. Indeed, the tariffs of the major trade partners are now so low that they have lost their protective effect in most cases (Pohl and Sorsa 1992: 13).

The EU is no exception. In Europe, the original six countries forming the Union completed the customs union in 1968, and members of the customs union increased to nine countries with the accession of the United Kingdom, Ireland and Denmark to the Union by 1973. 'Over the years since the formation of the European Union, rates of duty under the common external tariff have been progressively reduced in successive rounds of negotiations under the auspices of the GATT' (Hayes 1993: 11). At the beginning of 1986, a year ahead of the schedule originally agreed, the Union implemented the final stage of the tariff reductions negotiated in the Tokyo Round of multilateral trade negotiations which were completed in 1979.

The EU's tariffs are set out as conventional rates and autonomous rates (GATT 1991). The conventional rates generally apply to all imports from countries which are eligible for most-favoured-nation (MFN) tariff treatment. The countries include GATT contracting parties and several other countries for which MFN treatment has been agreed upon bilaterally. The autonomous rates or levies are applicable in those – relatively rare – cases where they are lower than the conventional rates or for categories for which no conventional rates are established. A wide range of food products falls into one of these categories.

Basic features of the EU's common customs tariff are as follows. First, the EU usually applies *ad valorem* duties, with relatively few exceptions.[13] Specific duties are imposed on some agricultural commodities (such as wine and spirits) which are one-third of all tariff lines as of 1992, and on coal[14] (GATT 1992c: 65; Pohl and Sorsa 1992: 13). In addition, a combination of specific and *ad valorem* schedules is applied to some fruits and vegetables and a few manufactured items such as glass and watches. According to Pohl and Sorsa (1992), the EU imposes specific duties on these goods mainly in order to facilitate customs valuation and administration. As an example, the EU insists that they impose specific duties on some goods in order to avoid customs delays resulting from valuation disputes about perishable goods.

Second, the EU applies low average tariffs. As of 1988, the overall tariff average was 7.3 per cent and 6.4 per cent for industrial products on a simple average basis, as shown in Table 6.4. As of 1988, about 29 per cent of items were free of duty, 25 per cent were charged at 0 to 5 per cent and 35 per cent at 5.1 to 10 per cent (GATT 1991). The remaining relatively high

Table 6.4 Tariff protection by sector (1988) (percentages)

Tariff Study category	*Average tariffs (simple)*	*Average tariffs (weighted)*	*Tariff range*
Textiles and clothing	10.1	7.6	0–17.0
Non-electric machinery	4.1	4.4	0–12.0*
Electrical machines and apparatus	5.8	8.3	0–15.0*
Foodstuffs	14.5	9.8	0–30.0
Cut flowers, plants, vegetable materials	8.4	4.4	0–24.0
Beverages and spirits	21.5	14.3	0–24.0
Ores and metals	5.1	2.6	0–17.0
Rubber	3.1	4.0	0–10.0
Tobacco	26.4	9.4	26.0–117.0
All industrial products (excl. petroleum)	6.4	5.6	

Source: *Trade Policy Review, European Communities* Vol. 1 (GATT 1993: 258, Table AIV. 1).
Note: '*' zero rates on certain products destined for use in civil aircraft. Weighted average tariffs were calculated using as weights the 1988 imports (under *ad valorem* tariffs) from MFN-treated countries. Tariff range refers to the range of conventional tariffs.

tariff rates – 12.5 per cent or more – are highly concentrated in a few categories of goods, although not all goods in these categories incur such high rates of duty.

The categories in question are agricultural products;[15] footwear; textiles and clothing; organic chemicals; some types of motor vehicles; some consumer electronics goods and electronics components. In fact, the Union doubled the tariff on compact disc players in 1984 to 19 per cent (NCC 1993: 95) and imposed 14 per cent tariff rates for most integrated circuits, radios, television receivers and VCRs after the full implementation of the Tokyo Round reductions in 1986 (GATT 1991: 89).

Third, not all European Union tariffs are 'bound' by GATT. A bound tariff cannot normally be increased (NCC 1993: 73). In particular, the GATT agreements of 1947 committed signatories to keep their tariff rates on industrial products fixed or 'bound', to reduce them in successive rounds of multilateral negotiations like the Tokyo Round and Uruguay Round, but never to raise them. As of 1993, the EU had a very high proportion of bound tariffs on industrial items. Almost 100 per cent of the Union industrial tariff lines[16] were bound, with about 65 per cent in agriculture.

Fourth, the EU has had significant tariff escalation in sectors such as fish, tobacco, leather, rubber, textiles and metals, as shown in Table 6.4. The average *ad valorem* tariffs for non-preferential imports of the six original Union countries were progressively reduced from 12.5 per cent in 1958 to 7.3 per cent on a simple basis in 1988, while the average *ad valorem* tariffs for fish, tobacco, leather, rubber, textiles and metals were substantially increased.

(i) *Preferential arrangements*

In the early days of the EU, the Union had a network of preferential tariff arrangements for Union trade in manufactured goods with the countries of the European Free Trade Association (EFTA),[17] with Mediterranean countries and with ACP countries under the successive Lomé Conventions[18] (Pohl and Sorsa 1992: 33).

In addition, a generalised system of preferences (GSP) for developing countries was agreed in the United Nations Conference on Trade and Development (UNCTAD) in 1971. 'The GSP was introduced in response to demands by developing countries for preferential access to help diversify their exports' (Pohl and Sorsa 1992: 34). Preferential agreements between the European Union and developing countries can be divided into three types: the Lomé Agreement, the Mediterranean Agreement and the GSP, as illustrated in Table 6.5. These preferential arrangements in the EU were made for political reasons. In other words, the EU has very complex trade relations with developing countries, because special relationships between some EU members and their former colonies or territories had to be considered in an EU-wide framework. In the relations with ACP, first, the residual feelings of obligation of France and the United Kingdom to their former colonial empires were reflected adequately[19] in the Lomé Conventions.

Second, special links with the Mediterranean countries were designed to promote stability in this neighbouring area. 'The twelve Mediterranean Agreements of association or cooperation are an expression of the historical

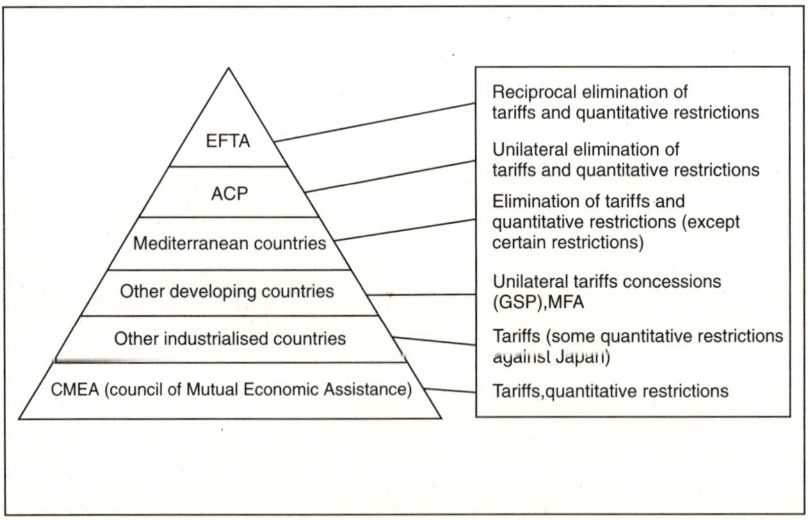

Figure 6.1 The pyramid of EU preferences for trade in manufactured goods
Source: Adopted from Möbius (1991)

interests and present political sensitivities of some EU members and also economic interests' (Pohl and Sorsa 1992: 33). Eighty-one developing countries in ACP and the Mediterranean have co-operation or association agreements with the European Union, providing for duty-free access for industrial goods and preferential access for some agricultural products, as Table 6.5 shows.

Third, the Union grants preferences on quotas to imports from developing countries, other than those covered by the Lomé Convention, under the GSP.[20] 'In fact, the GSP was introduced in 1971 in response to demands by developing countries for preferential access to help diversify their exports' (ibid.: 34). As Table 6.5 shows, the large majority of other developing countries (as of 1992, about 146 developing countries and territories excluding Eastern Europe and Taiwan), have relatively less preferential access – tariff rebates, duty-free quotas – under the GSP. In particular, a small number of very dynamic newly industrialized economies in East Asia which are graduating from the GSP, have also faced the brunt of EU safeguard actions and 'voluntary' restraints – mainly in textiles, but also in other 'sensitive' industrial goods. In addition, many agricultural products and some manufactures are excluded from the Union's GSP arrangements. For about 130 industrial products, there have been quantitative limits to tariff-free access under the scheme. As in the Union's other preferential arrangements, the value to the beneficiaries of the GSP scheme has been reduced by the progressive reductions in the common external tariff. The common customs tariffs can raise trade barriers in two respects. The first is that the GATT allowed countries[21] to raise tariffs unilaterally in two cases, which caused much anxiety, even though such tariffs have almost never been applied by the EU. Higher tariffs, in fact, have been applied by the EU twice by means of regulation No. 288/82. One cases was that Spain had invoked Article XIX to increase tariffs on synthetic rubber in February 1967. Spain started production in the Second World War under the Allied blockade.

Second is preferences on certain 'sensitive' agricultural and manufactured products which have been denied *ad initio*, or else imports were subject to quotas or limited by qualifying for preferential access. Sensitive products include certain chemicals, tyres, footwear, tableware, steel and electronics. These products are subject to country-specific limits on tariff rebates, differentiated according to import market shares in the European Union (6 per cent or 2 per cent of extra-EU imports). If limits are exceeded, duties are reimposed (Pohl and Sorsa 1992: 34). In addition, the EU has maintained high tariff rates in a few categories of goods.

The categories in question are agricultural products;[22] footwear; textiles and clothing; organic chemicals; some types of motor vehicles; some consumer electronics goods and electronics components. For example, the Union doubled the tariff on compact disc players in 1984 to 19 per cent and has imposed 14 per cent tariff rates for most integrated circuits, radios, television receivers and VCRs since the full implementation of the Tokyo Round

Table 6.5 EU trade preferences and their limitations for developing countries

Beneficiaries	*Agriculture*	*Industry*
ACP-LOMÉ AGREEMENT		
• 69 African, Caribbean or Pacific countries	• tariff quotas at reduced or zero rates with seasonal limits • quotas at reduced variable levies • sugar quotas at EU-supported prices for thirteen ACP countries • Banana quotas • national quotas	• duty free • some anti-dumping investigations
• Mediterranean countries	• Tariff quotas at reduced or zero rates with seasonal limits	• duty free
• 12 Mediterranean countries	• national quotas	• Voluntary export restraints on textiles, some antidumping measures, national quotas
GENERALISED SYSTEM OF PREFERENCES		
• LLDCs (8) • Andean group (4 Latin American countries) • Other developing countries (52 mainly Asian and Latin American countries, Eastern Europe, Baltic countries, Albania)	Four types of benefits: • reduced duties for fixed ECU amounts on sensitive goods • some non-sensitive goods duty free with seasonal limits • reduced variable levies for fixed amounts • some sensitive primary products excluded (metals and chemicals) • VERs, national quotas	• in general duty free, LLDCs and Andean exempt from quota limits Six types of limits to benefits: • sensitive industrial goods-fixed ECU limits (ceiling) by country • non-sensitive industrial goods – zero duties (China, Hong Kong, and Korea excluded), subject to safeguards • Multi-Fibre Arrangement (MFA) textiles-limited to signatories of MFA, LLDCs, and Andean; smaller tariff quotas for competitive suppliers; higher limits (ceilings) for others • other textiles – country-specific tariff quotas or limits (ceilings) • jute-zero duties for India, Thailand and LLDCs • steel-zero tariff quotas according to competitiveness • VERs, national quotas, anti-dumping measures

Source: *European Integration and Trade with the Developing World* (Pohl and Sorsa 1992: 72).
Note: LLDC = non-ACP least developed countries.

reductions in 1986. In evaluating the EU's GSP scheme, Brown (1988/9: 757) also concluded that 'the products of greatest immediate interest to the less developed countries have not been the object of significant reduction in tariffs or non-tariff barriers.'

The Union has also imposed quantitative limits to tariff-free access under the scheme on about 130 industrial products. These two restrictions also make non-EU member states believe that the EU still has a protectionist position in its trade policies.

6.2.3 Quantitative restrictions and restraint agreements with supplier countries

Thus, while rates of duty under the common external tariff were reduced, the emphasis on protection of manufacturing in the Union has subsequently shifted to quantitative restrictions and restraint agreements with exporters. The quantitative restrictions and restraint agreements are supplemented by anti-dumping and anti-subsidy measures.

In practice, there are four types of quantitative restrictions against imports from third countries, which member states continue to apply (Pohl and Sorsa 1992: 14). First, 'residual restrictions of member countries, predating the establishment of the European Economic Union, are applied to all third countries or to individual trading partners – but only by a member country. Prominent examples are motor vehicles and bananas.'

Second are 'a few EU-wide quantitative restrictions – agricultural products, textiles, and clothing under the Multi-fibre Arrangement (MFA), and footwear from the Republic of Korea and Taiwan (China)'.

Third, specific restrictions may be imposed by member countries on imports from state-trading countries, including China, Vietnam and North Korea.

Fourth, Portugal and Spain still impose a range of national quotas under the transitional provisions of their EU accession agreements. The national quotas of Portugal and Spain for industrial goods under the provisions were phased out on 31 December 1992. The national quotas of the two countries for agricultural products was also phased out in 1995. To some extent Greece also retained a number of trade restrictions as part of their transitional arrangements for joining the EU.

The most competitive Asian exporters, in particular the Asian NICs and China, have been the targets against which a few EU members have selectively used national quantitative restrictions. This is still going on. Most measures which the EU has taken are importer- and exporter-specific. The import volume affected by the measures of the EU covers a small share of Union imports from developing countries. As of August 1990, the member states, other than Spain and Portugal, restricted imports of more than 120 industrial products under Council Regulation No. 288/82 (at the four-digit

Table 6.6 Fixed duty-free amounts and ceilings of Korean consumer electronics products (as of 31 December 1990)

	Description	*Fixed duty-free amounts*	*Amount (ECU)*	*Ceiling (ECU)*
		Beneficiary countries or territories		
1	Television receivers (including video monitors and video projectors) which are not combined in the same housing, with radio broadcast receivers or sound or video recording or reproducing apparatus.	Korea* Singapore* Hong Kong*	650,000	4,410,000
	• Television projection equipment • Apparatus incorporating a video-phonic recorder or reproducer • Television receivers with integral tube	China	4,410,000	
2	Reception apparatus for radio telephony, radio telegraphy or radio broadcasting, whether or not combined in the same housing with recording or reproducing apparatus or a clock. Television receivers (including video monitors and video projectors) which are not combined in the same housing, with radio-broadcast receivers or sound or video recording or reproducing apparatus incorporating a video tuner and goods of category 1.	Korea* Singapore* Hong Kong*	650,000	4,410,000
3	Records, tapes and other recorded media for sound or other similarly recorded phenomena, including matrices and masters for the production of records but excluding products of Chapter 37 of Common Customs Tariff.	Korea*	3,000,000	9,450,000
4	Cathode-ray television picture tubes including video monitor cathode-ray tubes.	Korea*	800,000	2,646,000
5	Cathode-ray television picture tubes including video monitor cathode-ray tubes. • Black and white or other monochrome with a diagonal measurement of the screen not exceeding 52cm • Other cathode ray tubes	Korea	1,103,000	1,103,000

Source: *Official Journal of the European Communities* Nos. L 370/23 and L 370/24 (Commission of the European Communities 1990b).

Notes: Notwithstanding the rules for the interpretation of the combined nomenclature, the wording for the description of the products is to be considered as having no more than an indicative value, the preferential scheme being determined, within the context of the Annex of *Official Journal of the European Communities No. L370/7*, by coverage of the CN codes. (*) means the ceiling of fixed duty-free amounts of the countries will not change in 1991 because the items concerned are subject to the differentiation process.

tariff line level according to standard international trade classification, or SITC). As of August 1990, France (71 cases) and Italy (48 cases) accounted for the lion's share of the measures among more than 120 import restrictions on industrial products under Council Regulation No. 288/82. Most measures have been applied by France and Italy primarily in textiles to support MFA and other national quotas, and in footwear, consumer electronics and some agricultural goods.[23] Recently, the number of cases which the EU has restricted decreased significantly. In two steps taken in October 1991 and September 1992 (Council Regulations No. 2978/91[24] and 2875/92[25]) the number of restrictions was reduced to 30 for France and 19 for Italy.[26] The remaining measures focused on fruits and vegetables, including tomatoes, grapes, melons, apricots, pineapples and bananas, and on consumer electronics items, such as TV tubes, TV receivers and radio receivers of Japanese and other Asian origin. The Italians also restricted a variety of motor vehicle categories, as well as parts and components, imported from Japan; and France restricted imports of canned tuna and sardines.

Under the provisions of the Treaty of Rome (Article 33), member countries were allowed to maintain some quantitative restrictions on goods imported from non-EU member states provided they were in force when the treaty was signed. In 1968, following formation of the customs union of the six founder members, the Council of Ministers adopted a common list of products imported without quantitative restrictions from contracting non-EU parties to the GATT.

A general rule was adopted that member states would abstain from introducing quantitative restrictions on products figuring in the common list. If, however, there were marked increases of imports into the Union of any such products or of total imports in conditions causing or threatening serious injury to Union producers, the Council could decide by a weighted majority vote to withdraw products or exporting countries from the liberalisation list or, alternatively, to introduce Union quotas.[27]

In addition, an individual member country could institute emergency import restrictions, but such restrictions had to be notified immediately to the Commission for collective Union approval.[28] Between 1969 and 1974, the list of products on which the member states did not introduce quantitative restrictions was progressively extended.[29]

In addition, article 115 of the Treaty of Rome allows member states to block indirect imports by way of some Union countries with no restrictions, or less severe restraints than others, with the approval of the Commission. In fact, if it were possible to import via countries with no restrictions, or less severe restraints, quantitative restrictions in some Union countries with more severe restrictions could be circumvented.

The policy of blocking indirect imports was reconfirmed by the Maastricht Treaty on European Union which was signed in February 1992. The Maastricht Treaty continues the provision for the restraint of imports of particular items into individual member countries in order to

> ensure that the execution of measures of commercial policy taken in accordance with the Treaty by any member states is not obstructed by deflection of trade, or where differences between such measures lead to economic difficulties in one or more member states.
>
> (Pohl and Sorsa 1992: 26)

The problem, from the non-EU member states' perspective, lies in the fact that the restrictions have traditionally been applied at national rather than at EU level (GATT 1993: 67). Many EU import restrictions on industrial products are remnants of the member states' regimes before the founding of the Union. This was reconfirmed by the Council. 'In 1982, the Council finally agreed, "in a desire for simplicity and greater transparency of import arrangement," to draw up a list of quantitative restrictions still applicable at national level rather a common liberalization list'.[30]

This implies, in turn, that the restrictions are unsustainable within a fully integrated market. The internal market process should therefore result either in their abolition or in their harmonisation. Eventually, most national measures should disappear after the completion of the Single Market Programme in 1993, because of the likely enforcement difficulties in a frontierless European Union.

These measures are only a minor part of total tariff lines[31] and their restrictiveness varies greatly. The existence of Article 115 in the Treaty of Rome which restricts intra-EU transit of restricted third country exports, is a good indicator of the restrictiveness of the national measures (ibid.).

As of the end of 1991, the EU did not impose quantitative restrictions on any imports of Korean-made consumer electronics goods at Union level. However, France, one of the larger EU-member states, has imposed restrictions on the import of Korean-made consumer electronics goods (radios and CTVs) since 1984 through unilateral national quota. Quantitative restrictions on the imports of Korean consumer electronics goods have a small impact on Korean consumer electronics firms' decision to invest in production in the EU.

6.2.4 The New Commercial Policy Instrument and the common rules for imports (Regulation 288/82)

The European Union adopted Regulation no. 2641/84 which introduced the New Commercial Policy Instrument in 1984. Under the New Commercial Policy Instrument, the Union has also taken powers to take retaliatory action against non-Union countries if they are guilty or are thought to be guilty of illicit commercial practices. The Union can take retaliatory action by imposing any commercial policy measure compatible with the EU's international obligations and procedures if such action is considered to be in the Union interest, as a result of investigation. Where international obligations provide for consultation or for dispute settlement procedures, these must have been completed and their results taken into account before measures are decided (NCC 1993: 75–6). The powers which the Union have taken under the Instrument are similar to those given to the President of the United States by Section 301 of the US Trade Act of 1974. It is, however, specified that action should only be taken after failure to resolve the problem through the normal dispute settlement procedures. In fact, by mid-1991 the Instrument had been used in only two cases.[32] The purposes of these new powers are to 'defend vigorously the legitimate interests of the Union in the appropriate bodies, in particular GATT, and to make sure that the Union, in managing trade policies, acts with as much speed and efficiency as its trading partners' (NCC 1993:75). The regulation set up a procedural mechanism to respond to any illicit commercial practices and to remove the injury or to exercise the Union's rights with regard to third countries' commercial practices. In addition, Union regulations provide for remedial action in specific situations such as dumping, subsidisation or in laying the basis for trade-related policies, such as procurement and standardisation. These regulations are often related to EU obligations under GATT agreements.

Proceedings may be initiated either by an industry which considers itself injured or upon request from a member state. After consulting an advisory committee, the Commission decides whether to investigate or not. This decision is published in the *Official Journal of the European Communities*.

The Common Rules for Imports (Regulation 288/82) apply to all products, except agricultural products, textiles such as yarn and fabrics, and coal and steel.[33] These also regulate imports from all sources, except state-trading countries such as mainland China, North Korea and Vietnam. The common rules enumerate exceptions in an annex and allow maintenance of pre-existing national quantitative restrictions and surveillance measures by member states, even though the rules start from a general premiss of unrestricted market access.

6.3 EU policies towards the consumer electronics industry

Since the mid-1970s, European consumer electronics policy has been dominated by the politics of international trade (Cawson *et al.* 1990: 328, NCC 1993: 95). Specifically, the European Union has had a dilemma owing to a sharp conflict between industrial policy and competition policy.

According to its EU industrial policy, the Commission tried to force the pace of industrial collaboration between European firms through programmes such as the European Strategic Programme for Research and Development in Information Technology (ESPRIT) and Research and Development in Advanced Communications Techniques for Europe (RACE), although these programmes reflected an understanding of information technology which neglected the potential of consumer electronics. Although not directly responsible for EUREKA (the European Programme for High-technology Research and Development), the EEC supported the concept and tried to co-ordinate its own efforts with Eureka projects.

On the other hand, according to the competition policy, the Commission tried to prevent national governments from discriminating in favour of their national firms through subsidy and preferential public purchasing, and frowned on the unseemly scramble to attract Japanese investment. In addition, the External Relations Directorate was responsible for negotiations with Japan over trade friction and carried the responsibility for investigating allegations of dumping. In other words, it may be said that the European Community's electronics industry was protected from foreign competition by a variety of trade policy measures.

As discussed earlier, the EU drastically reduced tariff rates after the end of the Second World War, except on a few product categories. Consumer electronics goods was one of these categories. The Commission imposes customs tariffs of about 14 per cent for most integrated circuits, television receivers, radios and VCRs. Among consumer electronics products, the Union has offered less substantial tariff reductions to its trading partners on certain products. For example, the tariff on compact disc players was not cut back but doubled to 19 per cent in 1984. VER are frequently used. In mid-1988, 16-arrangements of various kinds limited imports of electronics products to the whole Union or to one or more of the member states. Japanese products covered include numerically controlled machine centres and lathes, CTVs and tubes, and VCRs, while Korean exports of VCRs were limited to the very low 1985 levels.

Until 1990, individual countries of the Union imposed various quantitative restrictions and restraint agreements on consumer electronics products imported from third countries. However, according to information available to the GATT secretariat, these arrangements were no longer in force in 1991 (GATT 1991). In addition, as has been seen, import restrictions are not uniform between Union countries, despite the efforts of the Commission to bring about a truly common external trade regime.

However, the main restrictions on market access take the form of non-tariff measures (NTMs). In particular, the instrument most favoured by the EU in recent years has been anti-dumping measures (NCC 1993: 96). In other words, 'the preferred policy instruments of the Commission shifted from voluntary restraint, through increased tariffs, to action on local content and stricter interpretation and enforcement of anti-dumping laws' (Cawson *et al.* 1990: 332). The major targets, as Table 7.3 (see next chapter) shows clearly, are Japan and Asian NICs, in particular Korea.

From 1983 to the end of 1990, the EU had initiated eleven anti-dumping investigations on electronics, as Table 6.7 shows. The products covered electronic typewriters from Japan and Korea (1984–8), microwave ovens from Korea, Japan and Singapore (1986), photocopiers from Japan (1986–9), videotapes from Hong Kong and Korea (1987–9), compact disc players from Japan and Korea (1987–90), VCRs from Japan and Korea (1988–9), audio cassettes from Japan and Korea (1988), dot-matrix computer printers and daisy-wheel computer printers from Japan (1988–9), small-screen CTVs from Korea, Hong Kong and China (1988–90) and car radios from Korea (1990). Of these, only in the case of microwave ovens was the anti-dumping measure not enforced. The other 10 all resulted in the imposition of duties or an acceptance of price undertakings, although in the case of small-screen CTVs[34] only provisional duties were in place.

In addition, imports of Japanese and Korean CTVs and VCRs are subject to voluntary restraint agreements and to EU surveillance. Further measures in the restraint of imports (import quotas, export restraint arrangements or monitoring procedures) are used by individual member states – France, Germany, Italy and Spain (Scott 1993: 384).

The Commission (April 1991) suggested that trade policy should be based on the following six objectives, in order to sustain a competitive Union electronics and IT industry.

- Maintenance of an open, multilateral international trade system;
- the improvement of access to the markets of the main trading partners in electronics and IT (notably the United States, Japan and South Korea);
- establishment of fair competition in international markets;
- support for scientific, technological, industrial and commercial co-operation in the international arena;
- continuing integration of European markets by means of new agreements with EFTA and East European countries;
- economic restructuring aid for the East European countries.

Despite the above suggestions, it cannot be denied that the European electronics industry is still protected from foreign competition by a variety of trade policy measures.

Table 6.7 EU anti-dumping cases involving consumer electronics imports since 1983

Year	*Product*	*Exporting country*	*Action*
1983	Video tape recorders	Japan	Investigation closed
1984	Electronic typewriters	Japan	Investigation opened/ provisional duties imposed
1985	Photocopiers	Japan	Investigation opened
1985	Electronic typewriters	Taiwan	Investigation opened
1985	Electronic typewriters	Japan	Definitive duties imposed
1986	Microwave ovens	Korea, Singapore, Japan	Investigation opened
1986	Photocopiers	Japan	Provisional duties imposed
1986	Electronic typewriters	Taiwan	Investigation closed
1987	Dot-matrix computer printers	Japan	Investigation opened
1987	Daisy-wheel computer printers	Japan	Investigation opened
1987	Compact disc players	Japan, Korea	Investigation opened
1987	Video cassette recorders	Korea, Japan	Investigation opened
1987	Video cassette tapes	Korea, Hong Kong	Investigation opened
1987	Photocopiers	Japan	Definitive duties
1988	Dot-matrix computer printers	Japan	Provisional duties imposed/ definitive duties imposed
1988	Audio cassette tapes	Japan, Korea	Provisional duties imposed
1988	Daisy-wheel computer printers	Japan	Provisional duties imposed
1988	Video cassette recorders	Korea, Japan	Provisional duties imposed
1988	Small-screen colour TVs	Korea, China, Hong Kong	Investigation opened
1988	Video cassette tapes	Korea, Hong Kong	Provisional duties imposed
1989	Daisy-wheel computer printers	Japan	Definitive duties imposed
1989	Video cassette recorders	Korea, Japan	Definitive duties imposed
1989	Video cassette tapes	Korea, Hong Kong	Definitive duties imposed
1989	Small-screen colour TVs	Korea	Provisional duties imposed
1989	Compact disc players	Japan, Korea	Provisional duties imposed
1990	Compact disc players	Japan, Korea	Definitive duties imposed
1990	Car radios	Korea	Definitive duties imposed

Source: *International Trade and the Consumer* (Grimwade 1990: 50–1).

6.4 The EU's policy-making mechanism on trade

Trade policy in the European Union is made in the framework of a written but ambiguous constitution laid down in the Treaty of Rome. The Treaty has certain problems. First, directives, which are a fairly strong instrument to implement common policies, have to be implemented in the national law of each EU member country, but national administrations have to comply with them even before their encoding in national law. The Treaty of Rome provides for different instruments to implement common policies (Pohl and Sorsa 1992: 12): regulations, directives, decisions, recommendations and opinions. Among these instruments, regulations are the strongest. Regulations are, in fact, directly applicable and legally binding in their entirety in all EU member states. Directives, as previously explained, are also a fairly strong instrument to implement common policies. The Commission supervises the implementation of common policies by members, issues formal notices of infringements and brings cases before the Court of Justice if infringements persist. The Commission has no direct means to enforce the decisions of the Court. The Commission relies on national enforcement. Union law is an integral part of national legal codes, and the same means of enforcement are available as for national laws. Enforcement is thus reasonably efficient – but not perfect – and varies across member countries. The Maastricht Treaty will strengthen the role of the Commission in Union law enforcement.[35]

Second, the thing with which non-EU countries are most concerned is that the Council has allowed significant discrepancies to remain between the approaches of different member states in every respect other than tariff rates (Hayes 1993: 33; Holmes 1991: 4). According to Article 113 of the Treaty of Rome, the Commission – which is the Union's central executive body – has powers of initiative in matters of trade policy and can negotiate with other countries on behalf of the Union as a whole. However, the European Council, comprising heads of government of member states and the Council of Ministers, has the powers of decision in matters of trade policy. 'In exercising the powers conferred upon it by this article, the Council shall act by a qualified majority'.[36]

In addition, external trade policy is also a Union matter, but the Treaty of Rome left the Council of Ministers much freedom to interpret the requirement for a true common commercial policy. The Treaty of Rome has also been interpreted by the Council of Ministers and the Commission with a degree of political expediency.

Hayes (1993) introduced four differing views on the nature of the decision-making process at the Union level. They are (1) inter-governmental politics, (2) functionalism and neo-functionalism, (3) transnationalism and (4) elite networks. The first view says that decisions on Union trade policy are made by negotiations between governments of member states. This interpretation raises questions at two levels: (i) how the desires and preferences expressed on behalf of the various member governments are arrived at and (ii) how agreement is reached on common Union policies.

The functionalist explanation is that policy-making in the Union is a problem-solving exercise in which common interests are stressed, giving scope to the transaction of business by technical experts. The 'neo-functionalism' explanation puts greater importance on self-interest as the prime motivation of political activity and sees interest groups, bureaucracies and political elites as competing at the Union level.

The third view is that the Union is part of a world-wide economic system of interdependence in which governments, international institutions and non-governmental groups become closely and directly dependent on one another. The holders of this view emphasise that some policy questions bring together particular groups of governmental and non-governmental participants, and policy decisions in some instances may be considerably influenced by Union-wide pressure groups or by coalitions of ministries from different countries. Transnationalism, like neo-functionalism, thus raises the question of whether governments are masters in their own houses. The last explanation is that decisions in the Union are often formed by particular coalitions of interest or by 'elite networks' consisting of limited numbers of individuals.[37]

The limitations of trade policy-making procedure in the EU lie in the fact that the Treaty provides no detailed rules for the conduct of trade policy. The governments, at least of the larger countries of the Union, do all in their power to maintain control of trade policy. In other words, it is very difficult to have a real common trade policy for all EU member states due to the conflicting positions of the larger countries of the Union. In fact, larger countries have been reluctant to recognise the powers of the Commission to take initiatives and to act for the Union as a whole in areas where the Treaty of Rome does not specify this clearly (Hayes 1993: 122). The conduct of trade policy has been left to subsequent Council negotiations and decisions. Larger member countries have taken full advantage of the provisions of Article 113 of the Treaty of Rome which lays down that the Commission shall conduct trade negotiations 'within the framework of such directives as the Council may issue to it'. This severely limits the freedom of action available to Commissioners and officials of the Commission.

The Treaty of Rome has formed the Article 113 Committee as an instrument to resolve the problem. The Article 113 Committee,[38] in fact, plays an important part in shaping Union trade policy. It is able to do so largely because the Union is made up of officials who know what is acceptable to the governments they represent. In fact, 'the committee meets weekly and, like the Committee of Permanent Representatives [COREPER], operates at two levels' (NCC 1993: 65). The full members, who are very senior government officials, meet once a month and are responsible for advising ministers on the Union's external policy objectives and strategy. The deputies also meet in the intervening weeks and deal with operational matters as well as preparing topics for the full members.

There are also arrangements, of varying degrees of strength, for inter-ministerial co-ordination of the handling of Union questions in the three largest member countries – Germany, France and the UK. If necessary, matters of trade policy are considered at the highest level of government. In addition, the various governments form their own views on matters of Union trade policy, considering what is likely to be acceptable to the representatives of other Union countries. However, there are certainly limits which cannot be transgressed unproblematically. This said, the governments concerned can fight vigorously when a major interest is involved.

6.5 Conclusion

Over the last decade the European Union has become more protectionist in a protectionist world, and this attitude has imposed significant costs on the EU's economy. However, the increasing protectionism of the EU was not the main reason for the increase of FDI of firms from non-EU member countries. The main reason was the uncertainty of the EU's external trade policy which affected exports of firms from non-EU member countries. Regrettably, this factor was ignored by conventional theories of FDI, such as Vernon's product life cycle theory. The trade diplomacy of the EU clashed with its competition policy principles as well.

'The external trade policy of the EU can still be characterized by instability, discrimination, lacking transparency and conceptual deficiencies' (Hiemenz *et al.* 1994: 56).[39] The characteristics of the EU's external trade policies can be summarised as follows. First, the EU's external trade policy concerns the preferred protectionist instruments, which reflect the EU's bias towards bilateralism, selectivity, discrimination and discretion. The most important NTBs continued to consist of VERs, EU surveillance and anti-dumping decisions (ibid.: 311). Not surprisingly, VERs and similar measures were concentrated on 'sensitive' sectors.

The EU has used various trade measures in order to protect its industries. Among these various measures, the importance of tariffs as one of the main instruments of protection has drastically declined. This is true of all industrial countries, partly as the result of post-war trade negotiations under the GATT and partly because of inflation and its effects on *ad valorem* duty. With the reductions of rates of duty under the common external tariff, the emphasis on protection of manufacturing in the Union has shifted to quantitative restrictions and restraint agreements with exporters. However, among these protection measures, anti-dumping duties have been used very frequently by the EU.

The EU has imposed anti-dumping duties on imports from non-EU member states, especially Japan and Asian NICs. For example, retrospective Union surveillance of imports has long been maintained *vis-à-vis* Japan and Korea on televisions and VCRs, and anti-dumping measures were, at mid-September 1992, in force against China and Hong Kong (CTVs and VCRs) and Korea (car radios, CD-players, CTVs, video cassettes and VCRs) (GATT April 1993: Vol. A, 161). Recently, the anti-dumping duties of the EU have caused much controversy, although the purpose of anti-dumping actions, at least, is to help ensure fair competition in the Union. Vague definitions of export prices and domestic prices and difficulties in proving the existence of injury or the threat of injury to Union producers with any certainty or objectivity, reveal the conceptual deficiencies of the EU's external trade policies. As examined earlier, the anti-dumping duties of the EU have been a cause of concern about 'Fortress Europe' for non EU-member countries.

Second, the EU's external trade policy relates to the traditionally large differences in protection levels across sectors and industries (Weiss *et al.*, 1988). The Union imposed moderate tariffs on a wide range of industrial raw materials, fuels and manufactured goods. However, high barriers (tariff rates) against non-European exports in 'sensitive' sectors (some chemicals, tyres, footwear, tableware, steel and electronics) remained largely unaltered.[40]

Third, the EU's external trade policies have the complex structure of preferential trade arrangements. The effects of preferences in the EU's trade relations with the beneficiaries have the inherent protectionist traps in many of the preference schemes (Hiemenz *et al.* 1990: 337; Langhammer and Sapir 1987; Sideri 1990). Pelkmans (1987: 38) expressed the viewpoint of less developed countries on the structure of the EU's preferential trade arrangements very well. He explained that 'serious criticism can be levelled at the discrepancy between lofty objectives and numerous impediments to free trade and security of access to the Union's market'. In fact, the schemes suffered from a serious lack of transparency, and the ensuing uncertainty added to the reluctance of Third World producers to invest (Langhammer and Sapir 1987).

Korean consumer electronics goods manufacturers invested in local production in the EU for the same reason. The exports of Korean consumer electronics firms have been restricted by the Union through many trade barriers and NTBs. As of the end of 1991, the exports of Korean consumer electronics firms to the EU were restricted by unilateral quotas by the French government (radio and CTVs) and anti-dumping duties at Union level. In particular, the imports of the Korean consumer electronics industry to the EU has been restricted mainly by the anti-dumping duties imposed by the EU.

The results of the interviews and questionnaire with senior managers of Korean consumer electronics firms in the EU showed that the EU's protectionist trade policies, not factors internal to the firms, were the strongest factor inducing investment by Korean consumer electronics firms in production within the EU.

Considering this, it should be concluded that Korean consumer electronics firms started investment in production within the EU in response to trade barriers and NTBs (mainly anti-dumping decisions) in the EU – a factor external to the firms.

7

INCENTIVE SYSTEMS OF EACH EU MEMBER STATE

As well as the trade policies and investment environments, various investment incentives offered by individual EU member governments are said to be a very important factor influencing the investment decisions of firms from non-EU member countries (including Korean firms) within the EU who see this as an alternative to exporting. The results of a survey conducted by the OECD (1989: 3), revealed that MNE-specific incentives have played an important role in influencing the location of FDI in particular countries, although they have not, in general, been a major determinant of the general decision to invest in the EU.

EU member countries then, have a problem in their investment incentive system. They do not have harmonised and standardised investment incentives for inward FDI of firms originating from non-EU member countries. The EU countries do their best to harmonise and standardise their investment incentives by following the EU's competition policy, because they know that this is very important for the achievement of the SEM. In particular, they try to harmonise and standardise investment incentives that are offered to foreign firms investing in poor regions.

However, the governments of individual EU member countries offer differing investment incentives in differing degrees to firms from non-EU countries, depending on the economic situation of each region. An example is the particularly attractive investment incentives offered to foreign firms (such as Samsung Electronics) investing in the Newcastle area in the UK. This is against the objectives of the EU's competition policy. Through the competition policy, the EU tries to prevent restriction or distortion of competition between member states, and the abuse of dominant positions by monopolistic firms, by regulating governmental support of each member state for inward FDI.

In the EU, in addition, concern has been expressed about excessive intervention by member states to attract investments (Gatsios and Seabright 1989). It can be said that the member countries have developed various investment incentives in order to promote domestic and foreign investment, especially FDI from third countries. The OECD has stressed that governments of EU member countries should stop competing with each other in offering investment incentives as it is considered to be a self-defeating process.

In this chapter, the common EU regulations on the incentive systems offered by member countries are reviewed, as well as individual countries' incentive systems. The aim is to analyse how the common EU regulations and various incentives offered by each member country affect the investment decision of non-EU investors to a particular country within the EU.

7.1 EU policies for government incentives of member countries

The EU supports through the common regional policy the projects of each member country which aim to correct structural and regional imbalances that might impede economic and monetary union. At the same time, as previously explained, through the competition policy, the EU Commission regulates governmental support of each member state for inward FDI in order to prevent, restrict or distort competition between member states and the abuse of dominant positions by monopolistic firms.

7.1.1 Common regional policy

At the Paris summit of 1972, the European Council had resolved to create a new agency – the European Regional Development Fund (ERDF) – in order to 'correct the structural and regional imbalances which might affect the realization of economic and monetary union' (Nevin 1990: 293). In 1973, Lord Thomson, the new Regional Commissioner, presented the Council with a report, *Proposals for a Community Regional Policy*, which became known as the Thomson Report. The bulk of this report was subsequently approved by the Council and made the first basis of a Union's regional policy. Eventually in 1975, ERDF was established. Much has been achieved by EU regional policy since then. The ERDF has undergone three major reforms – in 1979, 1984 and 1989 (Commission of the European Communities 1981, 1984c, 1989c).

Prior to the 1989 reforms, the ERDF had many problems. Armstrong (1993: 136) said that 'the history of the ERDF is largely one of a struggle to throw off the many restrictions imposed by the Council of Ministers in the original 1975 Fund Regulation'. There were no changes in many of the original restrictions on the ERDF's field of action immediately prior to the major 1989 reforms. In other words, the 1989 reform was a turning-point in EU regional policy. The 1989 reform changed most of the old structures and attempted to set up radical new principles to guide EU regional policy. Thus, the ERDF became an agency that could cope with the demands of the SEM and economic and monetary union (ibid.).

The reforms were started by two forces. The first was the Single European Act introduced in 1986. The decision to create the SEM, with the necessary legislation being set in place between 1989 and 1993, had major implications for EU regional policy. The second force was dissatisfaction with the existing ERDF. The 1989 reforms of the ERDF were conducted as part of a complete overhaul of all of the EU's Structural Funds. These comprise the ERDF, the European Social Fund (ESF) and the Guidance Section of the European Agricultural Guidance and Guarantee Fund (EAGGF). The aim of the overhaul was to ensure that the three Structural Funds acted in a closely co-ordinated manner and with a common set of objectives. Armstrong (ibid.: 140) pointed out that the three Structural Funds were set five priority objectives under the 1989 reforms (Commission of the European Communities 1988a):

1 promoting the development and structural adjustment of lagging regions (the ERDF, ESF and Guidance Section of EAGGF have responsibility);
2 the conversion of regions in industrial decline (the ERDF and ESF have responsibility);
3 the combating of long-term unemployment (the ESF has responsibility);
4 increased employment opportunities for young people (the ESF has responsibility);
5 (a) the adjustment of agricultural structures to accompany the reform of EU agricultural policy (the EAGGF has responsibility);
(b) the promotion of the development of rural areas (the ERDF, ESF and Guidance Section of EAGGF have responsibility).

In sum, EU regional policy has achieved much since the ERDF was established in 1975. The aim for the 1990s was to co-ordinate the new ERDF, which was made in 1989, with the ESF and EAGGF Guidance Section. However, although the EU's common regional policy is essentially a domestic matter for European states, which does not directly affect foreign firms from non-EU member states, access to these funds is nevertheless regarded as a factor which influences the investment decisions of foreign firms.

7.1.2 Competition policy

'The role and the enforcement of competition policy vary from country to country, and, in the same country, from decade to decade, reflecting changes in economic and social conditions as well as reversals in academic thought' (Jacquemin 1993).

Union competition rules cover two broad areas which are (1) enterprise conduct and (2) state intervention (Sapir, Buigues and Jacquemin 1993). The Treaty of Rome insists that competition rules should be applied to member states in so far as intervention creates or threatens to create specific distortions. In addition, EU competition policy applies to non-EU firms which sell or produce within the EU, but not to EU firms which engage in anti-competitive practices in export markets (Holmes 1991: 4). As Figure 7.1 illustrates,

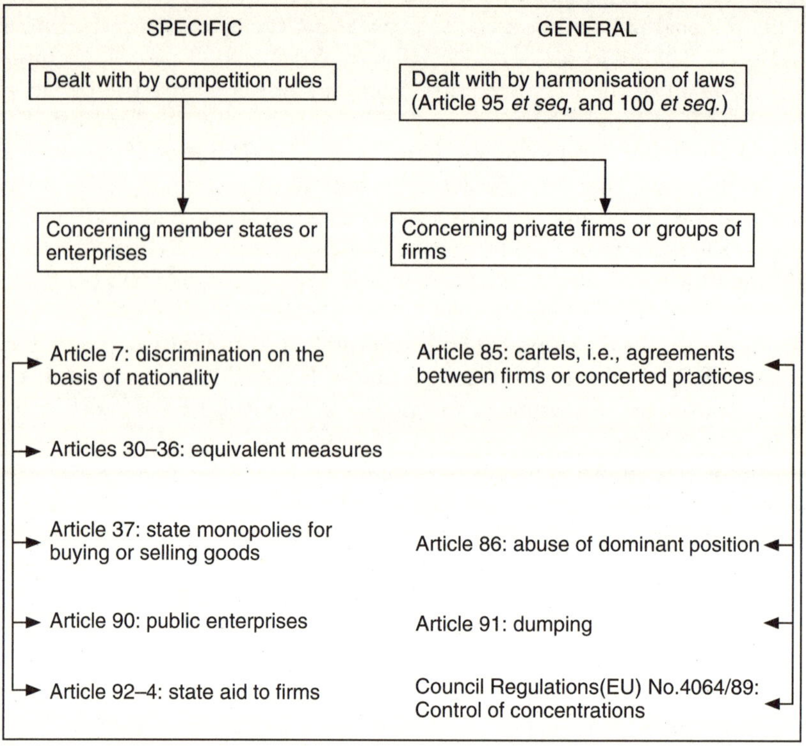

Figure 7.1 European anti-competitive policies: general and specific
Source: Henk Wouter De Jong (1993: 406)

rules concerning specific state intervention cover discrimination on the basis of nationality (Article 7), the conduct of public enterprises and state monopolies (Articles 37 and 90) and state aid (Article 92–4).

General distortions are supposed to be dealt with by harmonisation, as Figure 7.1 shows. General distortions are the result of differences between laws, regulations or administrative actions by member states. Rules regarding enterprise conduct include agreements (Article 85), abuse of dominant position (Article 86), mergers (Regulation 4064/89)[1] and dumping (Article 91).

'The EU uses its competition policy regulations in order to try to control and coordinate member states' regional policies' (Armstrong 1993: 147). Marques (1992) states that the EU regards its competition policy regulations as an important part of its regional policy. He goes on:

> In view of the less favoured regions' lack of financial resources and the limits to the resources of the Structural Funds (despite the recent

> doubling of those resources and the priority given to the less favoured regions), the effectiveness of aid in those regions depends to a very large extent on the action of competition policy. Only this policy is able to guarantee, through the monitoring of the various categories of aid in the central regions, that regional aid in the peripheral regions truly represents an advantage for those regions, capable of offsetting their structural weaknesses as far as attracting or generating investment is concerned.
>
> (Marques 1992: 407)

However, the EU cannot perfectly control governmental support of each member state for inward FDI through its competition policy. The EU Commission can control only the support offered by central governments of member countries. It cannot restrict the support offered by regional governments or local governments. Thus, Thies (1991) emphasises the necessity for stronger control by the EU through directives over the governmental supports of each EU member country for domestic or foreign investors.

7.2 Investment incentives of each EU member country

Generally, host countries offer the following investment incentives:

> (1) fiscal measures such as accelerated depreciation, preferential tax rates, tax exemptions and tax credits, and including also measures relating to social security contributions and investment reserves; (2) financial measures such as grants, preferential loans and loan guarantees and (3) other, non-financial measures including certain infrastructure-related assistance, preferential government contracts, the provision of certain services, and the establishment of free-trade, enterprise and technology zones.
>
> (OECD 1989: 11)

Among these three kinds of incentive measures, EU member countries employ, to varying degrees, five investment incentives in order to promote their objectives in a variety of policy areas such as regional development policy, industry and technology, service sector activities and small-firm development. These five investment incentives are capital grants, interest-related subsidies, tax concessions, depreciation allowances and labour-related subsidies, as Table 7.1 shows. Among these five types of investment incentives, capital grants remain the most important, although in Denmark and Greece the role of capital grants has been clearly reduced due to the abolition of the regional development grant in Denmark in 1991 and the switch of focus in Greece towards the fiscal aid package (comprising tax allowance and the increased depreciation allowance) in 1990.

Table 7.1 Distribution of regional incentives by country and incentive type

Country	*Incentive*	*Incentive types* (1)				
		CG	*IRS*	*TC*	*DA*	*LRS*
Belgium	interest subsidy		√			
	capital grant	√				
Denmark	(regional development grant)	√[a]				
	(municipality soft loan)		√[a]			
France	regional policy grant (PAT)	√[b]				√[b]
	local business tax concession			√		
Germany[c]	investment grant	√[e]			√[e]	
	special depreciation allowance				√	
	ERP regional soft loan[d]		√			
Greece	investment grant	√				
	interest rate subsidy		√			
	increased depreciation allowance				√	
	tax allowance			√[f]		
Ireland[g]	IDA grant — new	√				
	IDA grant — small	√				√
	IDA grant — services	√				√
Italy	capital grant	√				
	national fund scheme		√			
	social security concession					√
	tax concessions			√		
Luxembourg	capital interest/interest subsidy	√[h]				
	tax concession			√		
Netherlands	investment premium	√				
Portugal[i]	SIBR	√				√
	SIPE	√				
	SIFIT	√				√
Spain	regional investment grant	√				
UK						
• Great Britain	regional selective assistance	√[j]				
	regional enterprise grants[k]	√				
• N. Ireland	selective assistance	√[l]	√[l]			√[l]

Source: *European Regional Incentives* (Yuill *et al*. 1992: 431–2).

Note: Incentive type abbreviations:

CG: capital grant. *IRS*: interest-related subsidy. *TC*: tax concession. *DA*: depreciation allowance. *LRS*: labour-related subsidy.

IDA: Investment Development Authority. SIBR: System de Incentivos de Base Regional (Regional Incentive System). SIPE: Systema de Incentivos ao Potencial Endogeno (Financial Incentive System for the Development of the Business Environment). SIFIT: Sistema de Incentivos Financeiros ao Investimento no Turismo (Financial Incentive System for investment in Tourism).

a: These schemes were abolished as from 1 January 1991, though outstanding commitments will be met until 1993. Ad hoc support (by special Act of Parliament) may be available to potential inward investment projects which locate in the former Development Areas.

b: For manufacturing projects the PAT (Prime d'amenagement du Territoire) is generally an amount per job created up to a maximum expressed as a percentage of eligible investment. Given its investment-related ceiling, this component of the PAT scheme has been allocated to the 'CG' column in the table. For eligible non-manufacturing projects (basically research and tertiary activities), the PAT is calculated as an amount per job created. This component of the PAT scheme has been allocated to the 'LRS' column. The proposed new PAT-PP for small projects in rural areas takes the form of a capital grant.

c: A non-regional investment allowance (a capital grant) is available in the new eastern *Länder* from 1 July 1990 to 30 June 1992. The special depreciation allowance will be terminated by the end of 1994.

d: The ERP regional soft loan (ERP Regionaldarlehen) is a largely automatic, project-related concessionaire loan which can be awarded only to small or medium-sized firms for projects that are not eligible for the investment grant (i.e. So-called 'non-primary-effect' projects). The projects include projects of a basically local character including local services like wholesale and retail trades, craft activities and restaurants.

e: The special investment grant for high-grade jobs takes the form of a fixed amount per high-grade job created.

f: This is in terms of a fixed percentage of investment costs up to a maximum percentage of taxable profits in any given year. As a result of Law 1892/1990, the fiscal component of the Greek package has increased in importance.

g: Each IDA scheme consists of a package of assistance: new industry programme – includes capital grants, interest relief grants, loan guarantees, rent subsidies, training grants; small industries programme – includes capital grants, rent subsidies, employment grants, training grants; international services programme – includes capital grants, employment grants, training grants.

h: The overwhelming majority of awards have been in the form of a capital grant.

i: The SIBR has three components: grant-related 'industrial policy' and 'location' components and a job-related 'employment' component, the SIFIT has two components: a grant-related 'location' component and an award-per-job-created 'employment' component. The SIPE involves grant assistance.

j: Regional selective assistance normally takes the form of a capital grant but may also involve exchange risk cover for European loan.

k: These grants divide into investment grants and innovation grants and are for firms with fewer than 25 employees. l: The main elements of selective assistance are: industrial development grants, employment grants, interest-relief grants or soft loans, and marketing grants.

In Denmark the new liberal–conservative coalition government which was formed in December 1990 approved a new Industrial Development Act. According to this Act, regional development grants, loans and authorities were abolished with effect from 1 January 1991. Yuill *et al.* (1992: 131) found that the reason behind the considerable change came from the reordering of economic priorities in Denmark. According to them, the new Danish government attached more importance to the solution of national economic problems than to measures to reduce regional disparities, particularly when the unemployment rate in Copenhagen reached 10 per cent, which was equal to the national average. Budgetary problems also forced the Danish government to reduce public expenditure and grant aid to Danish industry which was already perceived as very effective in a European context. Finally, a reduction of industrial support measures was also pursued by the new government because they favoured solving economic development problems through non-interventionist policies.

Yuill *et al.* (ibid.: 47) also explained why the Greek government decided to enhance the attractiveness of the fiscal package. According to them, first, it was difficult for the Greek government to manage a highly discretionary system effectively because of the weaknesses of the Greek public administration and banking sector. The automatic tax allowance was one means of circumventing such difficulties. Second, the direct costs of the financial incentive package was considered to be excessive in the context of current budget deficits and thought to result partly from the abuse of the system by some investors, hence the more stringent monitoring procedures. Third, there had been political pressure from the private sector for greater liberalisation of the markets which were viewed to be 'suffocated by excessive administrative controls' (ibid.), of which discretionary incentive administration is thought to be a part. Fourth, there was seen to be 'a need to attract and retain investment (both Greek and foreign) in the light of the competition for mobile funds within the European Community, and increasingly from eastern Europe' (ibid.). In addition, incentives are open to abuse by officials in Greece. For example, Greek officials are said to take bribes from foreign firms which seek more attractive investment incentives from the Greek government.

Over the past decade, there has been a decreasing emphasis on loan-related subsidies and fiscal concessions[2] within regional incentive packages. As a consequence, capital grants became more important as investment incentives. In other words, capital grants have been an increasingly favoured incentive type within regional incentive packages due to their simplicity, flexibility and their compatibility with European Commission co-ordination requirements. Eleven EU member countries, but not Denmark which abolished capital grants as from 1 January 1991 because of its reorientation of economic priority, have capital grants within their incentive packages.

As of February 1992, only five EU member countries (Belgium, Germany, Greece, Italy and Northern Ireland) had interest-related subsidies.[3] Interest-related subsidies are no longer of major significance since general liberalisation of intra-EU exchange controls. In fact, the interest-related subsidies continued to decline in importance following the withdrawal of the municipality soft loan[4] in Denmark as from 1 January 1991 and the reduced role of the interest rate subsidy in Greece in July 1990. In addition, the fact that capital mobility means funds do not have to be local, also reduced the importance of interest-related subsidies as an investment incentive.

Fiscal concessions, then, have obviously increased in importance in Greece. On the other hand, Germany announced that the special depreciation allowance, available only in the Zonal Border Area, was terminated in 1994. Only five member countries will have regional fiscal incentives within their regional incentive packages.

Finally, as shown in Table 7.1, the labour-related subsidies were employed by six EU member countries as of the end of 1992. Even though these were used as a second favourable incentive type, they were generally low-key, with the notable exception of the Italian social security concessions to the Mezzogiorno.[5]

What is unique is that the importance of investment-related regional incentives on which the EU member states relied very heavily in the beginning of the 1980s has diminished because of the introduction of an interesting range of new measures. The EU member states have taken these new measures in order to develop the business environment in the EU more generally (Yuill *et al.* 1992: 8).

There has been much debate about a few issues concerning the investment incentives of the EU member states. One of the central topics for discussion is the extent to which regional incentives packages are automatic or discretionary in their administration. In the 1980s, there was a significant swing towards more discretionary support, with the demise of a number of large-scale automatic schemes, primarily due to budgetary pressures and a concern for 'value-for-money'. The large-scale automatic schemes that were dropped included the investment allowance in Germany and the regional development grant in Britain.

At the start of the 1980s, larger Union countries had an automatic policy measure for attracting FDI in their regional incentive packages coupled with discretionary 'topping up'. However, as of the end of 1992, most of the key regional incentives were essentially discretionary in their administration. As Table 7.2 shows, as of the end of 1992, fiscal concessions tended to be automatic, while most grants, in particular financial incentives generally, were subject to administrative discretion. Among five EU member countries which had financial incentives, four countries gave automatic incentive subsidies packages in administration, while Luxembourg had discretionary tax concessions.

Automatic schemes do offer certain advantages to investors. In particular, they are straightforward and relatively cheaper to administer than discretionary schemes. Associated with this, they do not involve civil servants in making commercial judgements. In addition, because of their predictability, automatic incentives can be incorporated into the investment decision-making processes of firms.

Automatic grants also have several problems. First, they can be wasteful, leading to windfall gains through support for projects which would have gone ahead even in the absence of financial incentives. In particular, 'in the case of very large projects the gains can be high and the budgetary impact significant'. Second, 'there is less contact, and therefore less scope for negotiation, between policy-makers and applicants. This may be important, for example, in the case of major inward investments where project content and issues such as local linkages can be influenced through discussions' (Yuill *et al.* 1992: 11).

Table 7.2 Degree of administrative discretion by country and incentive type

Country	*Incentive*	*Incentive types* (1)				
		CG	*IRS*	*TC*	*DA*	*LRS*
Belgium	interest subsidy		A/D[1]			
	capital grant	A/D[1]				
Denmark	(regional development grant)	(D)[2]				
	(municipality soft loan)		(D)[3]			
France	regional policy grant (PAT)	D				D
	local business tax concession			A		
Germany	investment grant	D				D
	special depreciation allowance				A	
	ERP regional soft loan		A			
Greece	investment grant	A/D[3]				
	interest rate subsidy		A/D[3]			
	increased depreciation allowance				A	
	tax allowance			A		
Ireland	IDA grant – new	D				
	IDA grant – small	D				D
	IDA grant – services	D				D
Italy	capital grant	A[4]				
	national fund scheme			A		
	social security concession					A
	tax concessions			A		
Luxembourg	capital interest/interest subsidy	D				
	tax concession			D		
Netherlands	investment premium	A/D[5]				
Portugal	SIBR	A/D[6]				A/D[6]
	SIPE	A/D[6]				
	SIFIT	A				A
Spain	regional investment grant	D[7]				
UK						
• Great Britain	regional selective assistance	D				
	regional enterprise grants	A				
• N. Ireland	selective assistance	D	D			D

Source: European Regional Incentives (Yuill *et al.* 1992: 435–6).

Note: Incentive type abbreviations.

CG: capital grant. *IRS*: interest-related subsidy. *TC*: tax concession. *DA*: depreciation allowance. *LRS*: labour-related subsidy.

IDA: Investment Development Authority. SIBR: System de Incentivos de Base Regional (Regional Incentive System). SIPE: Systema de Incentivos ao Potencial Endogeno (Financial Incentive System for the Development of the Business Environment). SIFIT: Sistema de Incentivos Financeiros ao Investimento no Turismo (Financial Incentive System for investment in Tourism).

D: Administrative discretion in award, rates up to a maximum. A: Little or no administrative discretion in award, rates fixed

1: In Wallonia, rates of award are determined by both quantitative criteria (under which awards are automatic, depending on job creation and value added per job) and qualitative criteria (under which awards are discretionary). In Flanders, there is discretion in the award decision (focusing mainly on project viability); thereafter, rates of award are largely predictable, though the (discretionary) 'strategic importance' component has recently increased from 3 per cent to 6 per cent of eligible expenditure.

2: These schemes were abolished from 1 January 1991, with outstanding commitments to be met until 1993. *Ad hoc* support (by special Act of Parliament) may be available to potential inward investment projects which locate in the former Development Area.

3: Although rates of award are now fixed, there remains discretion in the decision whether or not to make an award.

4: While the capital grant is basically automatic in character, the first and third implementation plans made provision for so-called 'planned bargaining', under which there is discretion to award the maximum possible capital grant (irrespective of project sector or location) to major industrial groups establishing innovative projects in the Mezzogiorno, large individual firms and certain consortia of small-and medium-sized firms.

5: For setting up projects with up to FI 18 million of eligible investment (FI 8 million for extensions), there is virtually no administrative discretion in the investment premium scheme. Rates of award are fixed and conditions of award are overt. In contrast, for setting up projects with more than FI 18 million (FI 8 million) of eligible investment. For setting up projects in excess FI 18 million (FI 30 million for extensions), awards in the three northern provinces in excess of the automatic maximum are wholly discretionary. In the transitional areas, the equivalent limits are FI 30 million for setting up projects and FI 75 million for extensions within five years of setting up.

6: The systems are overt. Applicants should know whether or not they will receive an award and also within what percentage range that award is likely to fall. There is, however, discretion regarding the actual percentage award made (except under the SIFIT).

7: Although basically discretionary, in certain zones a proportion of the award is determined automatically by employment created.

But automatic awards have played an important role in Greece following the switch of emphasis to the Greek fiscal aid package and the introduction of fixed rates of award under the financial aid package. However, the main 'automatic' country is Italy. As Table 7.2 shows, clearly, incentive packages that Italy has are automatically administered. Elsewhere, as Table 7.2 also shows, five countries have notable automatic components. They are Belgium, Greece, the Netherlands, Portugal and Great Britain. In Belgium, there is considerable predictability in rates of awards. In Greece, automatic awards now play a more important role than in the past as emphasis switched to the Greek fiscal aid package and the introduction of fixed rates of award under the financial aid package according to Law 1892/1990 in July 1990. In the Netherlands, awards are automatic for setting up projects with up to FI 18 million of eligible investment and for extension projects with up to FI 8 million. In fact, the Netherlands does not give its officials administrative discretion in the investment premium scheme. In Portugal, the system is overt and applicants know within what percentage range awards are likely to fall. Great Britain has an automatic administration system in regional enterprise grants which are restricted to firms with fewer than 25 employees. The regional enterprise grants were introduced on 1 April 1988 for firms in the Development Areas[6] with fewer than 25 employees. Such firms may qualify for investment grants of 15 per cent towards the costs of fixed assets subject to a maximum grant of £15,000 and innovation grants of 50 per cent to support product and process development subject to a maximum grant of £25,000. Because of their predictability, automatic incentives can be incorporated into the investment decision-making process of the firm.

On the whole, as Table 7.2 shows, many regional incentive packages are given with a large element of discretion in administration. In more detail, fiscal concessions tend to be automatic, while most grants, in particular financial incentives generally, are discretionary in administration. In particular, all the incentives packages in operation in which Ireland has are discretionary in administration.

Over time, there has tended to be relatively little change in those project types eligible for regional incentive support; they generally cover one or more of the following: setting up, extension, rationalisation, reorganisation, modernisation, spatial relocation, takeover, replacement.

As Table 7.3 makes clear, the main focus has remained on job-creating projects, such as setting up new factories or extending old ones, relocating plants from outside the problem regions. The support for job-saving projects such as rationalisation, reorganisation and modernisation has also been seen within most incentive packages, even though it has been subject to relatively stringent conditions. All EU member countries except for Denmark offer regional incentive support for setting up projects and extensions. Other incentive types are rarely given aid. Takeovers, for instance, tend to qualify for support only if most of the jobs 'at risk' are likely to be saved. France and Italy have regional incentive support for takeovers under complex and stringent conditions, as Table 7.3 illustrates. Replacement investment is almost everywhere excluded.

Amidst this relative stability in terms of project types covered, one unique trend which has drawn attention since the latter half of the 1980s has been the tendency to give preference to the setting up of new projects over extending old ones. This is because the setting up of new projects, in general, can create more jobs than extending old ones. Moves in this direction have been recorded in recent overviews in respect of Denmark, Germany, Ireland and the Netherlands. In the Netherlands, as an example, a new regional policy memorandum[7] put less emphasis on extensions and more on setting up projects. As Table 7.3 shows, some EU countries impose specific conditions before awarding regional incentive support for extensions. Ireland asks for a significant increase in existing capacity as a condition for awarding regional support for extensions. The Irish government offers regional grants only to new projects or major extensions which provide employment for more than 50 persons or investment of Ir. £ 500,000 in eligible assets. Germany also makes it a condition that the permanent labour force is increased by 15 per cent, or 50 new jobs.

Table 7.3 Project-type coverage

Country	*Incentive*	*Project-type coverage*				*Comments*
		SU/EX	*RR/M*	*SR*	*T*	
Belgium	IS, CG	√	√[1]	√[2]	X	• Project evaluation emphasises the employment effect of the Project and its regional development impact
Denmark	(RDG)[3]	√[3]	X	√[3]	X	
	(MSL)[3]	√[3]	X	√[3]	X	
France	PAT	√	X[5]	√[4,6]	√[7]	• At least 20 jobs (30 jobs for tertiary/research projects) must normally be created.
	IBTC	√	X[5]	√[4,6]	√[7]	• A complex matrix of job and investment conditions must be met.
Germany	IG	√	X[8]	√[9]	X	• Extension must increase the permanent labour force by 15% (or 50 jobs). This requirement does not apply to new firms extending within five years of setting-up.
	SDA	Not project-related				
	ERP	√	√	√[2]	X	• Extensions must create an appropriate number of new jobs.
Greece	IG, IRS, TA	√	√[10]	√[11]	X	• Projects must involve new productive investment for the purpose of establishing, expending, modernising or relocating units.
	IDA	Not project-related.				
Ireland	IDA (N – new),	√	X		X	• These grants form only part of the IDA aid package.
	(S – small)	X[12]				
	IDA (ISP – International Services Programme)	√	X[13]	X	X	• Extensions must represent a significant increase in existing capacity.
Italy	CG, NFSL	√	√[14]	√[2]	√[15]	• Specified eligible project types are: setting up, extension, modernisation, conversion, reconstruction, reactivation.
	SSC, TC[(16)]	Not project-related				
Luxembourg	CG/IS	√	√	X	X	• Specified eligible project types are: setting up, extension, conversion,

Table 7.3 (*contd.*)

Country	*Incentive*	*Project-type coverage*				*Comments*
		SU/EX	*RR/M*	*SR*	*T*	
						reorientation and rationalisation.
	TC	√	X	X	X	• The concession is available only in respect of new business and production lines or new high-technology-related services.
Netherlands	IPR	√[17]	√[18]	X	X	• Eligible projects need not include investment in a building.
Portugal	SIBR	√		√	√[19]	• There are no specific project-type criteria.
	SIPE	X[20]				
	SIFIT	Not relevant [21]				
		√	√	X	X	• Extensions and modernisations tend to be favoured.
Spain	RIG	√	√[22]	√	X	• SU/X should create new jobs. Modernisation and spatial relocation should maintain current employment levels.
Great Britain	RSA	√	√[23]	√[4]	X[24]	• (a) Basic distinction is between Category A (job-creating) and Category B (job-maintaining) projects.
	REG	√[25]	X[25]	√[25]	X[25]	• Eligible project types are not specifically identified.
N. Ireland	SA	√	√	√	√	• All projects must involve job creation or maintenance.

Source: Regional European Incentives (Yuill *et al.* 1992: 449–51).

Incentive abbreviations: IS: Interest subsidy. CG: Capital grant. RDG: Regional Development Grant. MSL: Municipality Soft Loan. PAT = Regional Policy Grant – Prime d'amenagement du Territoire. LIBTC: Local Business Tax Concession for Regional Development – Exoneration de la Taxe Professionnelle. IG: Investment Grant – Investitionszuschuss. SDA. Special Depreciation Allowance (Sonderabschreibungen nach Para. 3 des Zonenrandfoerderungsgesetzes). TA: Tax Allowance. IRS: Interest Rate Subsidy. IDA (Greece): Increased Depreciation Allowance; (Ireland) Investment Development Authority. SIBR: System de Incentivos dem for the Development of the Business Environment. SIFIT: Sistema de Incentivos Financeiros ao Investimento no Turismo (Financial Incentive System for Investment in Tourism). ISP: International Services Programme. NFSL: National Fund Soft Loan. SSC: Social Security Concession. TC: Tax Concession. IPR: Investment premium. RIG: Regional Investment Grant. RSA: Regional Selective Assistance. REG: Regional Enterprise Grants. SA: Selective Assistance

Project Type Abbreviations:
SU: setting up. *EX*: Extension. *RR*: Rationalisation and Reorganisation. *M*: Modernisation. *SR*: Spatial Relocation. *T*: Takeover.

Notes:

1: In Wallonia, the level of support for modernisation projects depends on the age of the equipment being replaced.

2: Only the *net* increase in investment is eligible (i.e. the value of the former settlement must be subtracted).

3: These schemes were abolished as from 1 January 1991. Eligible project types were: setting up of a new manufacturing plant or service firm (either indigenous or through relocation); indigenous firms beginning a new line of production of some scale; small firms undertaking a first major increase in production capacity.

4: Only projects moving *into* the designated region are eligible.

5: 'Internal conversion' is eligible. This involves a significant change of product (or products) requiring a major investment.

6: In practice, spatial relocation is restricted to moves away from the Paris region.

7: The establishment being taken over must be independent of the firm effecting the takeover and a 'significant' number of jobs must be maintained.

8: Conversion projects are only eligible in the new *Länder*.

9: Notes 2 and 4 apply. In addition, if *Länder* boundaries are crossed the 'sending' Land must give permission before aid can be paid.

10: RR is eligible only to the extent that they are related to the application of modern technology.

11: Enterprises moving to a worse-off region (or to a designated industrial estate) receive a relocation supplement under the IG/IRS schemes.

12: While not eligible under the main IDA grant programmes, takeover may qualify in the context of IDA support which aims to facilitate mergers and takeovers.

13: Eligible only if they involve a shift of headquarters or research functions to Ireland from abroad. Spatial relocation within Ireland is not eligible.

14: Modernisation projects are eligible, i.e. projects which, within existing employment limits, increase productivity or improve working conditions or improve ecological conditions relating to the production process. Under the 1986 Mezzogiorno law, conversion and reconstruction projects are also eligible.

15: Reactivation is eligible i.e. the takeover of a bankrupt plant.

16: The ILOR reinvestment concession is project-related, with eligible project types being broadly the same as (but, if anything, slightly wider than) under the CG/NFSL schemes.

17: Extensions must increase productive capacity or employment by at least 20 per cent. Moreover, this increase must relate to *all* the activities of a company (or a group of companies) within a range of 10 kilometres. This requirement may be waived or reduced in the case of 'footloose' projects involving more than FI 30 million investment.

18: Restructuring projects must improve the regional economic structure by introducing new production methods or new products or by involving new markets. This is a stringent award condition.

19: The move must be from a non-designated to a designated area.

20: Takeover is eligible only in exceptional cases and then only with the approval of the Minister of Industry.

21: The scheme assists intangible investment: sectoral studies; economic, financial and technical feasibility studies; technical and management consultancy; and capital market studies.

22: Rationalisation and reorganisation are not eligible.

Table 7.3 (contd.)

23: Category B projects must safeguard a 'substantial number' of jobs; involve some action to prevent job loss; and bring about a fundamental rationalisation or modernisation outside the normal run of capital expenditure which, moreover, achieves a significant and identifiable improvement in job protection.
24: Eligible if qualifying as a Category B project or, exceptionally, a rescue case.
25: Although eligible project types are not specifically identified, proposed projects are expected to be viable and to have a positive effect on sales, profits and employment levels.

In addition, there are several other characteristics of the EU regional incentive packages in the 1980s and 1990s that would affect the investment plans of some Korean firms. First, in the 1980s a number of EU member countries decentralised regional incentive administration, at least for small projects and firms. This trend has been reinforced by recent developments in France and Spain.

In France, a proposal has been adopted to introduce the PAT-Petits Projects (PAT-PP), a new regional incentive for smaller projects in rural areas. Although in many ways similar to the Prime d'amenagement du Territoire (PAT), or Regional Policy Grant, the proposed new scheme differs in respect of two important eligibility conditions. First, it will have a minimum investment target of FF 2 million, while it is FF 20 million under the PAT. Second, it will be restricted to PAT areas which are classified as Objective 5b Regions, which means rural areas in need of development.

Another key difference between the new scheme and the PAT is that the PAT-PP is administered at the regional level. This reintroduces a regional component to the administration of the French regional aid package, which had previously been lost in 1987 when the regional tier of the PAT scheme was withdrawn due to budgetary pressures.

In Spain, budgetary pressures underlie the administration of the main regional incentive, the regional investment grant. One of the steps taken to reduce expenditure concerns has been to consider restricting the grant to projects involving investments exceeding Pts 75 million. It means that Spanish regions provide support for projects below the Pts 75 million limit from their own budgets.

Second, in the 1980s and early 1990s, only two trends have been worthy of note with regard to activity coverage in which four countries (the Francophone region of Flanders in Belgium, Greece, Italy and Portugal) recorded significant changes. First of all, there has been a move to extend coverage towards industrial services. The second trend has been a tendency to increase the level of rate discrimination between eligible industries. What is interesting is the fact that industrial characteristics have come to play a more central role in award decision-making.

> In this context, it is of note that the countries concerned all retain a significant automatic component to their regional incentive awards. The more automatic are awards, the greater the emphasis on explicit rate discrimination in award, not least in order to ensure that incentive spending remains within budgetary limits.
>
> (Yuill *et al.* 1992: 21)

Third, in respect of spatial coverage of the EU regional incentive packages, the major trend of the 1980 was a significant and ongoing reduction in the spatial coverage of designated problem regions, especially in the northern Union countries. The trend continued strongly into the 1990s, in part as a response to continuing European Commission pressures and in part as a reflection of ongoing budgetary constraints. Three countries, Denmark, Germany and the Netherlands, recorded major cutbacks for problem regions. Italy also recorded major problem region cutbacks even though a less significant reduction was shown. Belgium, France and Luxembourg are currently reviewing problem region boundaries.

Yuill *et al.* (ibid.) make two key points. First, 'the cumulative effect of the cutbacks in the problem region coverage which have taken place in the (mainly northern) Community countries since the early eighties has been major'. Indeed, the 'norm' in terms of population coverage now lies at well under 30 per cent of their population (compared with two-fifths previously) in designated problem regions. The second point is that they feel that EU involvement in the various developments has been remarkable and is likely to continue.

Fourth, in respect of size coverage, minimum size conditions have been introduced in schemes in a number of countries such as Belgium (the Francophone Flanders), Greece and Spain, primarily due to budget constraints. In Belgium, the minimum investment limit was increased in May 1990 from BF 10 million to BF 20 million. The Belgian government should reduce its budget because its economy has worsened and foreign debt has increased to the point of exceeding GNP by a considerable margin since 1974 (Maldague, Leroux and Gilot 1993: 134). In Greece, minimum investment targets under both the investment grant and interest rate subsidy schemes were introduced by the government (Law 1892/1990). These targets vary by sector and project type. The minimum investment required for setting up and extension projects in manufacturing sector was Drs 60 million. Since the second oil shock, the Greek economy has suffered stagflation and high fiscal imbalances, as well as occasional balance of payments crises. The reasons for this include a fast-expanding and inefficient state sector, the increasing regulation of market forces and unsustainable income demands (Economou, Tsaveas and Politis 1993: 267). In Spain, the budget also had to be cut because its economy was in the crisis until 1986 and such serious problems as weak uncompetitive industries, despite the capacity for growth, have dogged the country since (Sanchez 1993: 402).

However, changes in size coverage among the main incentives on offer in the Union countries in the 1980s have tended to be few in number and minor in significance. The main trend in the 1980s, if it can be called a trend, has involved an increase in the minimum size of projects for which grants are made. The minimum investment thresholds have been raised in order to reduce the number of projects and to focus on the more significant ones. However, this trend has been balanced in some countries by the introduction of specific schemes for small firms and projects. In general, size restrictions do not play a major role in regional incentive systems in Europe. In practice, only France has size minima of note (certainly in terms of job creation), while just a handful of regional schemes are explicitly targeted at small and medium-sized firms.

Fifth, in the 1980s, overall nominal rate maxima in the Union countries have, in general, not changed markedly, because they has been determined primarily by European Commission aid ceilings. However, within these overall maxima, there have been moves to introduce lower award ceilings for particular activities, areas, project types and group sizes, especially in those countries where there has been major legislation (Denmark, Greece and the Netherlands). The changes have often been in response to budgetary pressures. However, in the 1990s, the level of change in respect of maximum nominal rates of award has indeed been significant so that it has affected all EU member countries except France, Luxembourg and Spain. In considering the change in the level of incentives to foreign-owned firms (FOFs), it is useful to divide EU member states into three groups.

The first group includes Denmark, Greece and the Netherlands, where the changes reflect major new laws on regional incentives. In Denmark, regional incentives were abolished from 1 January 1991, although the government has reserved the right to offer a grant of up to 25 per cent of eligible expenditure to potential major inward investment projects which settle in former Development Areas. This compares with the previous regional incentive system which offered regional development grants at a maximum 35 per cent in Priority Regions and 25 per cent in Ordinary Regions. In Greece, Law 1892/1990 has resulted in a notable shift in terms of rate maxima from the financial to the fiscal regional aid package with, for instance, a cut of 10 percentage points in the nominal rate maximum of the investment grant. In the Netherlands, there have also been rate cutbacks. The New Regional Policy Memorandum has reduced the rate of award for the investment premium from 25 per cent to 20 per cent of eligible expenditure for setting up projects and extensions within five years of setting up, and to 15 per cent for extensions more than five years after setting up.

The second group includes Belgium, Germany and Northern Ireland, where the changes have accompanied important new policy initiatives. In the Francophone region of Flanders in Belgium, the discrimination between eligible activities has been increased.

> Three groups of eligible sectors have been created: priority sector (maximum basic rate; 8 per cent); intermediate sectors (standard basic rate: 6 per cent); and other eligible sectors (no basic award, but may qualify under other award components). Previously, the first two groups qualified for the higher basis award (12 per cent) while the third group received the standard basic award (8 per cent).
>
> (Yuill *et al.* 1992: 459)

In addition, the conditions governing acquisition of the incentives have been strengthened and the countries have got more discretion in giving incentives to FOFs under the 'strategic importance' aspect of policy. In fact, 'the maximum award for projects of "strategic importance" has been increased from 3 per cent to 6 per cent of eligible expenditure' (ibid.). However, the overall award ceiling remains unchanged.

In Germany, the new eastern Länder have the same general award ceilings that previously applied in the Zonal Border Area, which is a maximum 23 per cent for setting up projects and 20 per cent for extensions. However, the new eastern Länder has a rate maxima of 15 per cent for rationalisation and reorganisation projects as well as for conversions. Previously, the rate for these was 10 per cent. Following the ending of the Zonal Border Area preference, these ceilings compare with rate maxima outside the new Länder of 18 per cent, 15 per cent, 12 per cent or 10 per cent depending on location and project type.

In Northern Ireland, three distinct rate maxima which were varying by area have been replaced by a single 30 per cent maximum. However, this was combined with a supplementary discretionary award of up to 20 per cent, taking the overall award maximum back to the 50 per cent figure which previously applied.

Considered together, the important new regional policy developments which have taken place in all three countries have seen significant changes in the determination of award rates for specific projects in specific locations. However, as a whole, the same overall rate maxima for the regional incentive package has been retained.

The third group includes the Republic of Ireland, Italy and Portugal, where the composition of the regional incentive package is largely unchanged but where there have been noteworthy developments in terms of nominal rates of award. In the Republic of Ireland, the rate maximum for extensions under the new industry programme has been reduced further from 25 per cent to 15 per cent as from 1 January 1991. In Italy, the extension of priority sector status in Region A to include virtually all eligible industries has resulted in a further favouring of Region A over the other designated parts of the

Table 7.4 Maximum rates of capital grant award by country

Country	*Designated problem region*	*Population coverage*[1]	*Max. capital grant award (%)*[2]	*Max. EU aid ceiling (%)*[3]
Belgium	Zone 1	22.0	20 nge	20 nge
	Zone 2	11.1	15 nge	15 nge
Denmark	Priority Regions	11.1	25[5]	25/20 nge
	Ordinary Regions[4]	8.6	25[5]	17 nge
France	Maximum Rate Zone	39	25	25
	Standard Award Zone	39	17	17
Germany	East Germany Länder	21	23/20/15	23/20/15
	GA Area	27.0[6]	18/15/12/10	18/15/12/10
Greece	Thrace	12	45–55	75 nge
	Other Region D	12	35–45	75 nge
	Region C	32	25–40	75 nge
	Region B	14	15–40	75 nge
	Region A	42	0–40	75 nge
Ireland	Designated Areas	34	60	75 nge
	Non-Designated Areas	66	45	75 nge
Italy	Mezzogiorno – Backward	8.9	56	75 nge
	– Intermediate	21.7	48	75 nge
	– Other	5.0	40	75 nge[7]
Luxembourg	Designated Areas	n.a.	25/20/17.5	25/20/17.5
Netherlands	IPR Areas	19.9 (8.0)	20	20net/25/15
Portugal	Zone 3	100	75	75 nge
	Zone 2	100	45	75 nge
	Zone 1	100	33	75 nge
Spain	Zones of Industrial Decline			
	Zones of Economic Promotion	4.1	75/45/30/20	75/45/30/20 nge
UK		54.2	50/40/30/0	50/40/30/0 nge
• Great Britain	Development Areas	15.5	30 nge	30 nge
	Intermediate Areas	19.5	20 nge	20 nge
• N. Ireland	Whole Province	100	50	75 nge

Source: *Regional European Incentives* (Yuill *et al*. 1992: 459, Table 4.1).

Notes:

1: Population coverage of designated problem regions as a percentage of the national population.

2: Maximum nominal rate of capital grant award. In both Belgium and Britain the rate is expressed in terms of the EU aid ceiling, i.e., in net grant equivalent terms, after tax.

3: Nge = net grant equivalent after tax. The full list of EU ceilings can be found in the *Official Journal* 12 August 1988 (OJ No. C 212).

4: That is, those Ordinary Development Regions which are not designated as Priority Development Regions.

5: The Danish regional incentive package was abolished from 1 January 1991. However, the government has reserved the right to offer a maximum 25 per cent grant to major inward

investment projects locating in the former Development Regions.
6: The GA area figure is as a percentage of the West German population. It represents just over one-fifth of the total national population.
7: From the end of 1990 the nge ceiling for Frosinone fell to 30 per cent. In addition, other parts of the Mezzogiorno lost their designated status at the end of 1990 or 1992.

Mezzogiorno. In Portugal, the determinants of the industrial policy component of any award has changed, with the generation of foreign exchange no longer being taken into account. Along with a location component and an employment component, the industrial policy, however, has become a key determinant in the decision of whether or not to make an award.

Table 7.4 underlines the importance of rate discrimination by area. It also shows that the Union countries fall into two broad groups with respect to nominal rate maxima. The first group includes Spain, Portugal, the Republic of Ireland, Northern Ireland, Italy and Greece where the maximum in question is 50 per cent or more.

The second group includes Denmark, Britain, the Benelux countries, France and Germany, where nominal maxima are generally only about half this level. In addition, Table 7.4 makes it clear that, by and large, these groupings reflect European Commission aid ceilings. The EU set its aid ceilings based on 'the general principles governing all types of aid and on two principles peculiar to it alone – geographical concentration and modulation of the intensity of aid' (Marques 1992: 404). In the first group of countries the Commission ceiling is generally 75 per cent net grant equivalent after tax, while in the second it is 30 per cent net grant equivalent or less. The EU Commission controls, as explained previously, the governmental support of EU member countries by setting the maximum aid ceiling. It means that EU member countries can support any industries within these maximum rates. However, if any EU member countries support an industry more than European Commission aid ceilings or bring any threat of distorting the competition among member countries, the Commission can control the countries by using the Competition Law.

7.3 Influence of incentives system of each EU member state on the investment decisions of Korean consumer electronics firms

As the research by Root and Ahmed (1978), Lim (1983), O'Sullivan (1985), Guisinger and Associates (1985), Guisinger (1986), Balasubramanyam (1991) and OECD (1989) revealed, MNE-specific incentives offered by the EU have also played some role in influencing the location of FDI of Korean consumer electronics firms within the EU, although they have not, in general, been a major determinant of such investment. In addition, as Siggel (1986) insisted, the interviews with nine Korean consumer electronics firms operating in the EU reveal that incentives give the greatest impact on investment decisions of Korean consumer electronics firms when they are undecided about their choice of a number of economically viable locations and when the incentives are not in conflict with performance requirements.

For this research, a survey was conducted with 26 Korean consumer electronics firms in the EU.[8] Sixteen firms out of 26 answered the question on

Table 7.5 Korean consumer electronics firms' top five reasons for locating in a particular country within the EU

	A	*B*	*C*	*D*	*E*	*F*
UK						
Kolon Imperial Graphics P.L.C.	2	3	4	4	3	1
SKM Europe, Ltd.	1	1	1	2	–	3
Inkek (UK) Ltd.	4	4	4	3	3	3
GoldStar Electric (UK) Ltd.	4	5	4	5		
Trigem Computer (UK) Ltd.	1	3	3	4	1	4
Samsung Electronics (UK)	2	2	1	2	2	1
Daewoo Electronics UK Ltd.	4	3	2	4	3	5
Average (UK)	**2.6**	**3**	**2.7**	**3.7**	**2.3**	**3.4**
Germany						
GoldStar Europe GmbH	3	4	3	2	3	3
Samsung Electronische Bauelemente	3	4	3	–	4	1
Samsung Corning Deutschland GmbH	5	2	4	1	2	4
Average (Germany)	**3.7**	**3.3**	**3.3**	**1**	**3**	**2.7**
Ireland						
Saehan Media Ireland Ltd.	5	3	4	3	2	1
SaRonix Kony Ireland Ltd.	5	4	4	3	3	1
Average (Ireland)	**5**	**3.5**	**4**	**3**	**2.5**	**1**
France						
Haitai Europe S.A	5	3	5	1	4	4
Italy						
GoldStar Iberna Italy SPA	5	1	1	2	1	3
Portugal						
Samsung Portugal Produtos Electronecanicos S.A	4	1	1	3	1	1
Spain						
Samsung Electronica Espanola S.A.	1	4	2	4	4	1
Average (All)	**3.38**	**2.94**	**2.88**	**2.87**	**2.67**	**2.56**

Notes:
A: Offer of government grants at more favourable conditions than other EU countries.
B: Better quality of labour.
C: Favourable investment enviornment in terms of infrastructure, government efficiency or political stability.
D: Cheaper labour than other EU countries.
E: Operation of a number of other Korean manufacturers.
F: Comparatively larger market size than other EU markets.

the choice of a particular country as the place of local production within the EU, as Table 7.5 illustrates. Of the 16 responding firms, seven were in the UK (including one firm in Northern Ireland), three in Germany, two in the Republic of Ireland, and one in each Italy, France, Portugal and Spain. The answers of these 16 firms overall were that the offer of government capital grants at more favourable conditions was the major determinant of locating in a particular country in the EU. Better quality of labour, they said, was the second major determinant. However, the answers of the seven responding firms which started local production in the UK differed somewhat from the norm of the 16 responding firms, as Table 7.5 shows.

The major reason for locating in the UK was that it had cheaper labour costs than other EU member countries. The second major reason was the UK's comparatively large market size. The third was the better quality of labour in the UK than in some other EU member countries. None of these top three reasons for locating in the UK were related to the incentives system offered by the UK government. Indeed, the UK government does not offer a labour-related subsidy to foreign investment in the UK in the regional incentives packages. The UK government offers labour-related subsidies to foreign investment only in Northern Ireland.

Clearly, then, offers by the UK government are not less attractive than those of other EU member countries. The UK government offers three incentive types (capital grants, interest-related subsidies and tax concessions) to firms from non-EU member countries, as Table 7.1 shows. These are not less attractive incentives in comparison with Germany, which offers four incentive types (capital grants, interest-related subsidies, depreciation allowances and labour-related subsidies), Greece, which offers four incentive types (capital grants, interest-related subsidies, tax concessions and depreciation allowances), or Italy, which offers four incentive types (capital grants, interest-related subsidies, tax concessions and labour-related subsidies). In project-type coverage, the UK does not have a less attractive package than other EU member countries either, as Table 7.3 illustrates. Along with all other EU member countries, the UK also offers incentives to the setting up of new projects. In particular, the UK government offers incentives to all types of projects (setting up, extension of old ones, rationalisation and reorganisation, modernisation, spatial relocating and takeover), as long as the projects involve job creation or maintenance.

The UK has received the number of FDIs by Korean consumer electronics firms (10 cases) of all EU member countries. From this, it is open to doubt whether incentive systems play an important role in influencing the location of Korean consumer electronics firms in a particular country within the EU. However, the incentives are not the major determinant in Korean firms' decisions in relation to location, size or timing of investments.

The answers of three responding firms which started local production in Germany showed three major reasons for locating in Germany. They were, the important position of Germany in the world market, the offer of more favourable tax incentives in Germany and the offer of German government grants at more favourable conditions than other EU countries. These three firms stated that the incentive system offered by the German government was their major determinant for locating in Germany.

The answers of two responding firms which invested in new plants in the Republic of Ireland showed similar results to the overall answers of 16 responding firms. The major determinants of locating in Ireland were said to be the offer of government grants at more favourable conditions than other EU countries and easier access to EU restructuring funds.

The interviews with senior managers of nine responding Korean consumer electronics firms also revealed that the incentive system offered by the EU was a very important factor, but not a major determinant when they made decisions in relation to location, size or timing of their investment within the EU.

In sum, the answers to the questionnaire revealed that the incentive system offered by the EU was certainly a very important factor which had a large influence on the location of Korean consumer electronics firms in a particular country in the EU. But the incentives, in general, were not the major determinant of location.

7.4 Conclusion

The governments of each EU member country offer differing investment incentives to different degrees to firms from outside the EU. The various governmental incentives of each EU member state are controlled and supported by the EU Commission, directly or indirectly. In other words, they must offer the volume of investment incentives within the European Commission aid ceiling. On top of this, each EU member government offers some very attractive investment incentives which are more than the European aid ceiling, especially in poor regions, in order to attract investments from firms originating outside the EU. This is certainly against the objectives of the EU's competition policy, and could have negative effects on the achievement of the SEM. The competition policy objective is to prevent restriction or distortion of competition between member states and the abuse of dominant positions by monopolistic firms.

The main reason why the EU cannot address this problem lies in the fact that the EU cannot perfectly control governmental support of each member country for inward FDI through its competition regulations. While the EU can control support offered by the central governments of member countries, it can only imperfectly control support offered by regional governments or local governments.

In the 1980s and early 1990s, there were changes in the investment incentives offered by the EU. In the 1980s, first, there was a significant swing towards more discretionary support mainly due to budgetary pressures and a concern for value-for-money. In terms of project-type coverage, since the latter half of the 1980s, the first priority has been put on job-creating projects such as the setting up of new projects or extending old factories, or relocating plants from outside the problem regions. With regard to activity coverage, there was a move to extend coverage towards industrial services and to increase the level of rate discrimination between eligible industries. There was also a significant and ongoing reduction in the spatial coverage of designated problem regions, especially in the northern Union countries in the 1980s and 1990s. Even though the changes were not significant, generally the investment incentives offered by the EU became less favourable to investors coming from non-EU countries.

The number of Korean consumer electronics firms that started local production within the individual EU member states was contingent not only on the number of incentive types offered by each EU member state. The firms said that, certainly, investment incentives had an influence on decisions related to location, size and timing of investments, but they were not generally viewed as the key determinants of overall investment levels. As discussed in Chapter 8, the results of the survey conducted with the 26 Korean consumer electronics firms in the EU proved that investment incentives, especially capital grants, had a stronger influence on decisions relating to location of investment. The results of interviews with senior managers of nine Korean consumer electronics firms also revealed that they thought the investment incentives offered by the EU affected strongly the size and timing of their investments in a particular country within the EU.

8

THE FDI OF KOREAN CONSUMER ELECTRONICS FIRMS IN THE EU

Empirical survey

The first example of Korean FDI was a project in forestry development in Indonesia in 1968. This was typical of the labour-intensive and resource-oriented investment that has been made in Third World countries such as China, Indonesia, Malaysia, the Philippines, Thailand, etc. This, then, is the first type of Korean FDI. The second type is FDI in advanced countries. The first type has increased more than the second both in value and number of FDI projects, as Table 8.1 shows. Since 1988, in particular, Korean FDI in China has recorded the highest growth rate both in value and number of FDI projects of all Korean FDIs. In terms of value, as shown in Table 8.1, Korean FDI in China, based on the number of projects approved, recorded a 8,442 per cent increase during the six years from 1989 to 1994. In terms of number of projects, Korean FDI in China, again based on number of approved projects also recorded an increase of 6,967 per cent during the same period.

The second type of FDI, investment in developed countries (especially in the USA and EU) has been primarily in capital- and technology-intensive industries (Park 1985: 29). This pattern of Korean foreign investment is very similar to that of Japan.

Since 1986, as analysed in Chapter 2, Korean FDI in the EU has increased steadily, although the growth rate has not been as high as that in East Asian countries and the USA. Most Korean FDIs in the EU have been made by consumer electronics firms. Most of these firms belong to a *chaebol*. The relationship between a *chaebol* and firms belonging to it can be regarded as paternalistic, because a *chaebol* is a group of firms owned or controlled by a person or family.

As of the end of 1994, 26 out of 46 cases of Korean manufacturing FDI in the EU had been made by Korean consumer electronics firms. However, two out of the 46 cases of FDI in the EU[1] cannot be regarded strictly as FDI since they involved small amounts of money (about US$ 5,000) and were not

Table 8.1 Korean FDI in China, Thailand, Malaysia, Indonesia and the Philippines

		1989		*1990*		*1991*		*1994*	
		No. of FDI projects	*Value US$ 1,000*	*No. of FDI projects*	*Value US$ 1,000*	*No. of FDI projects*	*Value US$ 1,000*	*No. of FDI projects*	*Value US$ 1,000*
China	A	12	9,770	38	54,493	112	84,721	1,067	824,772
	B	7	6,360	23	15,974	69	42,468	836	640,806
Thailand	A	15	15,452	30	41,651	13	32,631	19	25,874
	B	9	9,141	16	12,807	16	32,489	14	28,119
Malaysia	A	18	35,588	31	70,898	26	32,445	21	170,036
	B	7	3,081	18	17,529	27	69,555	16	20,120
Indonesia	A	57	135,135	98	400,205	49	198,850	21	66,309
	B	41	75,390	60	163,979	57	170,075	21	67,865
The Philippines	A	15	11,839	23	70,774	45	39,168	274	91,634
	B	9	2,645	16	32,147	27	48,269	166	45,019
USA	A	72	272,995	106	473,632	96	402,494	140	723,261
	B	66	168,520	80	342,627	84	395,244	128	530,794
Japan	A	28	12,912	24	16,483	18	113,191	25	227,412
	B	21	10,179	15	10,925	21	14,934	16	57,952
Canada	A	10	162,455	6	122,098	2	118,445	10	108,241
	B	6	114,399	6	92,335	4	68,066	9	42,490

Source: *HaeWae TuJa TongKae YunBo* [Overseas Direct Investment Statistics Yearbook] (Bank of Korea 1992 and 1995).

Notes: A: Approval basis. B: Invested basis.

intended to start up as ongoing operations. Of the 26 cases of FDI made by Korean consumer electronics firms in the EU, 20 were made by companies belonging to one of the *chaebols*.[2] Of these, 17 companies belonged to three major *chaebols* (Samsung, nine companies; Lucky-GoldStar and Daewoo, four companies each).

Generally, however, one should be very careful in analysing the power of multinationals on the basis of statistics. As Stopford, Strange and Henley (1991: 14) point out, statistics may not fully demonstrate how powerful the multinationals are. In the case of three Korean multinationals (Samsung, Lucky-GoldStar and Daewoo), however, it may be said that the statistics on FDI prove that they are really major companies because they possess a majority of the equity capital of all Korean companies operating in the EU. They took 26.55 per cent (US$ 24,370,000) of the total value of Korean FDI in manufacturing sector in the EU between 1985 and 1992 (US$ 91,728,000).[3]

There are many possible reasons for the advance of the Korean consumer electronics companies into the EU. To discover which of these reasons was decisive, I conducted a questionnaire survey of the European operations of 26 Korean consumer electronics companies in 1995. To get more in-depth information, I carried out, also in 1995, a series of interviews with senior managers of nine Korean consumer electronics firms operating in the EU.[4]

The remainder of this chapter is in three sections. First, profiles of the three major *chaebols* (Samsung, Daewoo and Lucky-GoldStar); Second, a review of the findings of my questionnaire survey and interviews with Korean managers; and third, a conclusion.

8.1 Profiles of the three major Korean *chaebols* to which consumer electronics companies in the EU belong

8.1.1 Samsung

Samsung is the largest *chaebol* in Korea. In *Fortune*'s (25 July 1994) list of the top 500 international industrial corporations, as at the end of 1993, Samsung was the 14th-largest industrial corporation in the world, with total sales of US$ 51,345.2 million. Samsung is also a highly diversified and adaptive enterprise that competes effectively in the global economy Samsung's business can be divided into seven distinct fields: trade; electronics; heavy industries and chemicals; precision instruments; food processing and textiles; service; and culture and welfare, as Table 8.2 shows (Steers, Shin and Ungson 1989: 50). Since the death of its chairman, Lee Byung-Chull, Samsung has been run by the dynamic Mr Lee Kun-Hee, the third son of the late chairman.

The Samsung group was established by late chairman Lee Byung-Chull in 1936. Mr Lee started as a provincial businessman under the Japanese colonial regime. He established a local rice mill with funds inherited from his parents

Table 8.2 Principal business fields and companies of the Samsung Group (as of 31 December 1994)

Business field and company	*Main products or services*
Trade	
Samsung Co. Ltd	Exports, imports, natural resource development
Electronics	
Samsung Electronics Co. Ltd	Colour and black-and-white televisions, VCRs, stereo systems, home appliances, industrial electronic equipment
Samsung Electron Devices Co. Ltd	Picture tubes, CRTs for data display, display monitors, receiving tubes, transistors, diodes, computers
Samsung Electro-Mechanics Co. Ltd	VHF and UHF tuners, condensers, ceramic capacitors, speakers, deck mechanisms
Samsung Corning Co. Ltd	Television glass bulbs, electronic gun-mounts, glass rods, glass bulbs for computer display
Samsung Semiconductor and Telecommunications Co. Ltd KwangJu Electronics Co. Ltd	64K DRAM, 256K DRAM, and 1 MB DRAM semiconductors, optical cable and fibres, electronic switching systems, computers, facsimiles
Heavy Industry and Chemicals	
Samsung Shipbuilding and Heavy Industries Co. Ltd	Industrial machinery and equipment, structure steel, shipbuilding and offshore structures
Samsung Petrochemical Co. Ltd	Purified Terephthalic Acid
Samsung Construction Co. Ltd	Civil engineering, architectural construction and plant installation
Samsung Engineering Co. Ltd	Plant engineering
Precision Instruments	
Samsung Aerospace Industries Ltd	Jet-engines, electro-optics equipment, cameras, industrial robots, process control systems, computer peripherals
Samsung Watch Co. Ltd	Digital and analogue watches, electronic display boards
Food Processing and Textiles	
Cheil Sugar & Co. Ltd	Sugar, MSG, flour, IMP, GMP, cut pork, formulated feed, soybean meal, vegetable oil, processed meat, premix organic fertilisers
Cheil Wool Textile Co. Ltd	Worsted textiles, woollen textiles, men and women's garments, knitwear, blankets
Cheil Synthetic Textiles Co. Ltd	Polyester staple fibre, polyester filament
Services	
Samsung Life Insurance Co. Ltd	Individual and group life insurance

Table 8.2 (contd.)

Business field and company	*Main products or services*
Samsung Fire and Marine Insurance Co. Ltd	Fire, marine, and causality insurance
Chosun Hotel	Hotel services
Hotel Shilla Co. Ltd	Hotel services
Shinsegae Department Store Co. Ltd	Consumer retail
Shinsegae DaeJeon Station Store Co. Ltd	Consumer retail
Samsung Card Co. Ltd	Credit card business
HaeWonDae Development Co. Ltd	
Samsung Securities Co. Ltd	Securities brokerage, underwriting dealer operation
Samsung Investment and Finance Corporation	
YeunPo Leisure Development Co. Ltd	
Culture and Social Welfare	
Joong-ang Development Co. Ltd	Development of groundwater resources
Joong-ang Daily News Co. Ltd	Daily newspaper, magazines

Source: Adopted from *Samsung: Its Role and Activities as a General Trading Company*: 1986–1987 (Seoul, 1987): 46–7, *The Chaebols* (Steers, Shin and Ungson 1989: 52–3); *1996 Nyun 30 Dae KiEup GipDan* [The Analysis of the 30 Korean Big Business Groups for 1996] (Choi 1996: 154–5).

Note: MSG: monosodium glutamate.

(Jones and Sakong 1980: 352). Owing to the success of the rice mill, by 1945 he had expanded into brewing, trucking, real estate, domestic trading,[5] milling and noodle-making. These activities were largely confined to Taegu, one of the five big cities in Korea, located in the southern area of the Korean peninsula.

After liberalisation from Japan in 1945, Mr Lee became interested in starting a new line of business in Seoul, using funds accumulated primarily from his brewery profits. He moved to Seoul in 1947 and decided to engage in international trade. He established the Samsung Mulsan [General Trading] Company in Seoul in November 1948. This venture was devastated by the Korean War, but was reestablished in Pusan, the second-largest city in Korea, in January 1951. Mr Lee made significant profits in the import trade before, during and immediately after the Korean War. The success of this company laid the foundation for the *chaebol*'s subsequent rapid expansion.

The foundation established in the early 1950s was further strengthened through the mid-1950s. In this period he moved into import substitution of consumer goods such as sugar, flour and textiles, and from there into a wide range of services and products. In the late 1950s Samsung emerged as a formidable *chaebol* and continued taking over numerous firms, including financial intermediaries. In particular, Samsung purchased almost half of all commercial bank shares in the later years of President Rhee's regime.[6] Samsung's success can be explained by Mr Lee's management style, which was influenced by the Japanese style of management. As well as beginning his entrepreneurial activities under the Japanese colonial regime, he was educated for awhile in the Political Science Economics Division of Waseda University, failing to graduate from that university due to poor health. In fact, along with the GoldStar Group, it is arguable that the Samsung Group maintains a Japanese style of management. Samsung adopted OJT (on-the-job training), QC (quality control), a limited form of life-time employment, centralised decision-making and the practice of placing great emphasis on the 'family symbol' of the company (Park 1985: 30).

Throughout the 1960s, Samsung continued to expand into new fields such as department stores, securities brokerage, educational institutions, real estate, paper mills and mass media. In the early 1970s, in particular, the group moved into electronics and, in the mid-1970s, it entered the 'heavy and chemical industries' emphasised by the nation's Fourth Five-Year Economic Development Plan. The Samsung Group established or acquired companies in petrochemicals, heavy equipment, precision machinery, construction and shipbuilding (Jones and Sakong 1980: 353). This story is in contrasts quite to most business histories in Japan, where it was much more difficult for small new companies to grow by diversification due to the legal and cultural impediments to M&A.[7]

Samsung Electronics Co. plays an important role in the business activities of the Samsung Group. The Samsung Electronics Co. has contributed much to the Samsung Group's ambitious plan to become a truly international corporation.

> In 1982 it opened a television assembly facility in Portugal with a capacity of 300,000 units for sale within the European common market. This was followed in 1984 with a US$ 25 million plant in New Jersey that produces 1 million televisions and about 400,000 microwave ovens per year. In 1987 another US$ 25 million facility was opened in England with a capacity for 300,000 VCRs, 400,000 color televisions, and 300,000 microwave ovens.
>
> (Steers, Shin and Ungson 1989: 30)

The company's goal was to double its exports from Korea (from US$ 2.2 billion to US$ 5 billion) by the early 1990s, becoming one of the top seven electronics manufacturers in the world (*Business Korea* 1988). As of the end of 1993, in fact, Samsung Electronics Co. ranked fourth in *Fortune*'s list of the 44 largest electronics and electrical equipment corporations in the world, with total sales of US$ 51,345 million, while it was fifth in 1992 with total sales of US$ 49,560 million.

8.1.2 Daewoo

The Daewoo (which means 'great universe' in Korean) Group was in all probability the fastest-growing *chaebol* of the 1970s. It was the third-largest *chaebol* in Korea in terms of value-added as of the end of 1992. Today the Daewoo Group is involved in various businesses as Table 8.3 illustrates. In 1975 Daewoo was only the seventh-largest *chaebol* in Korea in terms of value-added. The Daewoo Group has several unique features distinguishing it from other *chaebols*. First, it was in all probability the fastest-growing *chaebol* of the 1970s. Second, the Daewoo Group is one of the 'new generation' of business groups (Jones and Sakong 1980: 361). Mr Kim Woo-Choong, founder of the Daewoo Group, was 8 years old when Korea gained independence and did not start his business until the high-growth years of the 1960s.

The Daewoo Group was founded in 1967 as a small textile trading company with an investment of US$ 18,000 and four employees. In 1967 Mr Kim started business with three of his close business friends to form their own trading company – the Daewoo Industrial Company Ltd. They got initial success from this export trade. Owing to this success, Mr Kim and his friends expanded their business into export production. In the mid-1970s, they diversified into a variety of areas including construction, electronics and heavy machinery (ibid.).

To diversify their business, the Daewoo Group initiated a series of 14 acquisitions. The strategy was the same for each acquisition, and each time it was successful; Daewoo took over an ailing company, improved the management system and product quality, and turned the company around. One of the hallmarks of Daewoo has been its superior production capability. The first major takeover occurred in 1975, when Korean President Park asked Daewoo to acquire a state-owned machinery plant that had been losing money for 37 years.

Kim Woo-Choong, founder of Daewoo, changed the name of the company to Daewoo Heavy Industries and took personal control of the new company. Within one year, the company broke even and by the second year it began paying dividends. Shortly thereafter, President Park also asked Mr Kim to take over Okpo Shipbuilding Company. The shipbuilding company is now Daewoo Shipbuilding and Heavy Machinery Ltd, which is one of the world's largest and most respected shipbuilding companies. After take over of this shipbuilding business, Kim invested US$ 500 million in completing the facility. Finally, Daewoo took over Saehan Motor Company in 1979. This company is today Daewoo Motor. Again, through Kim's efforts, the enterprise was transformed into one of Korea's two largest auto companies.

In this way, the Daewoo Group has adopted a three-pronged strategy of company creation, company acquisition and joint venture. Through this

Table 8.3 Principal business fields and companies of the Daewoo Group (as of 31 December 1994)

Company	*Business field and products*
Trading and Construction	
Daewoo Corporation	Trading: export, import, international finance, resource development
Keangnam Enterprises, Ltd	Plants, architectural and civil works
Hotel and Finance	
Dongwoo Development Co. Ltd	Seoul Hilton Hotel
Daewoo Securities Co. Ltd	Securities brokerage, underwriting dealer operation
Daewoo Research Institute	Economic and survey
Daewoo Investment and Finance Corporation	Operation of short-term Cp, dealer operation in money markets
Heavy Sector	
Daewoo Heavy Industries Ltd	Diesel engines, rolling stock, construction equipment, industrial vehicles, machine tooling, aerospace
Daewoo Shipbuilding and Heavy Machinery Ltd	Ships, offshore structures, medium-sized vessels, yachts
Daewoo–Sikorsky	Helicopters
Electric and Electronics Sector	
Daewoo Electric Motor Industries Ltd	Motors for electric products
Orion Electric Co. Ltd	TV tubes, monitors, VCR heads
Daewoo Electronics Co. Ltd	TVs, VCRs, PCs, audio products, home applicances
Daewoo Electronic Components Co. Ltd	TV tuner, capacitor
Orion Electric Components Co. Ltd	Electric components
Daewoo Telecom Co. Ltd	Telecom switching system, PCs fibre optics, O/A machines, semiconductors
Daewoo Information System Co. Ltd	Information systems
Automotive Sector	
Daewoo Motor Co. Ltd	Passenger cars, buses, trucks, speciality vehicles
Daewoo Automotive Components Co. Ltd	Cranking motors, alternators, ignition distributors, ignition coils, gears, axles, brakes, radiators
Koram Plastics Co. Ltd	Polyurethane bumpers
Daewoo Precision Industries Ltd	Auto parts and components

Source: *HanKoo KyungJaeeue KulRurk LiDong* [The Shift of Power in the Korean Economy] (Kong 1995: 336)

strategy, Daewoo has continued to record about a 20 per cent annual growth rate in output and to overtake its global competitors. In 1993, according to *Fortune* (25 July 1994), Daewoo ranked 33rd in its list of the 500 largest industrial corporations in the world, with total sales of US$ 30,893 billion; whereas it had ranked 39th in 1986 and 41st in 1992. In 1993 Daewoo also ranked 91st in total profits, 22nd in total assets, 54th in stockholders' equity and 81st in total employees.

As with the Hyundai Group, the Daewoo Group used a management method which may be classified as a Western type[8] (Park 1985: 30–1). In fact, Daewoo has been dominated largely by young and highly efficient senior managers. Its managerial practice is very similar to that of Hyundai, particularly with regard to the advance into new investment sectors (e.g. car plants) and countries (e.g. Canada, the People's Republic of China).

Daewoo Electronics Co. Ltd founded in 1974[9] as a manufacturer of small cassettes is now a major member of Korea's dynamic Daewoo Group. As an integrated manufacturer of electric and electronic products, the company now produces everything from electronics and home appliances to computers, components, telecommunications and other high-tech products, as Figure 8.1 illustrates. According to the company, total sales in 1983 of US$ 220 million rose to US$ 416 million in 1985, over US$ 1 billion in 1987 and a stunning US$ 1.6 billion in 1988, constituting a high annual growth rate over the period. With total sales for 1991 at over US$ 2.5 billion, Daewoo Electronics Co. Ltd had indeed established itself as a reputable international manufacturer of quality products. It now has over 16,000 employees including over 2,000 professional engineers engaged in research. By establishing a solid reputation for quality products, Daewoo Electronics has seen an increase in its capital from US$ 48 million in 1985 to US$ 428 million in 1991, thereby providing plenty of power for further growth. In 1993 Daewoo Electronics Co. Ltd

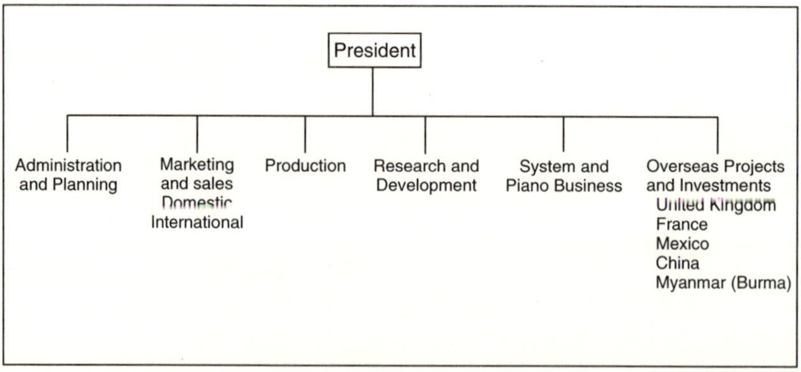

Figure 8.1 Organization of Daewoo Electronics Co. Ltd.

ranked ninth in the world in electronics and electrical equipment (*Fortune* 25 July 1994).

Daewoo Electronics utilises the extensive Daewoo world-wide network of more than 80 branch offices to gather information on the latest consumer needs and trends, and in 1988 these efforts resulted in exports of over US$ 1 billion worth of quality goods. This was about a ten fold increase in just five years over the 1983 export figure of US$ 92 million.

8.1.3 Lucky-GoldStar

As of the end of 1995, the group consisted of 48 affiliated companies and was managed by the dynamic Chairman Koo Bon-Mu. Sales for 1995 totalled US$ 52 billion, while employment stood at an estimated 105,476 (Choi 1996: 124). Among the companies belonging to LG group, LG International and LG Electronics were ranked 304th and 340th in the world, respectively, in terms of total revenue in the 1995 Fortune International 500 (*Fortune* August 1996).

Lucky-GoldStar was founded by Koo In-Whoi in 1947. He established the Lucky Chemical Company (later Lucky Ltd) which initially manufactured cosmetics (especially facial creams). He later expanded his business into plastics (including combs and toothbrushes). After the Korean War, he opened a new small-size company specialising in consumer goods (such as dental brushes, dental cream and soap) in 1953 using the profits made by the Lucky Chemical Company before the war. Owing to business success, in 1958 he also set up GoldStar Co. Ltd to produce radios, refrigerators (1965) and televisions (1966). An oil refinery was established in 1967. From here, the company has continued to grow as a 'dual-track' entity, stressing product innovations in both chemistry (through Lucky Ltd) and electronics (through GoldStar Co. Ltd). Later the Group also started services and public welfare businesses.

Like Samsung, the Lucky-GoldStar Group has committed large amounts of resources to applied research. Lucky-GoldStar spent US$ 348 million in 1986 and US$ 470 million in 1987 on R&D. This represents 5 per cent of total sales (compared with 3 per cent for similar companies), and plans are in place to increase this to 7 per cent by 1998.

The company maintains numerous research labs, but its major efforts have been centred in its central research institute in Taedok, which was established in 1979, and in a major R&D complex in Anyang, which was established in 1985. The Central Research Institute in Taedok has carried out research in chemical technologies and advanced material sciences, while the R&D complex in Anyang has focused on electronics research.

GoldStar has been also aggressive in overseas production. In 1982 the company opened a manufacturing plant in Huntsville, Alabama, which has since

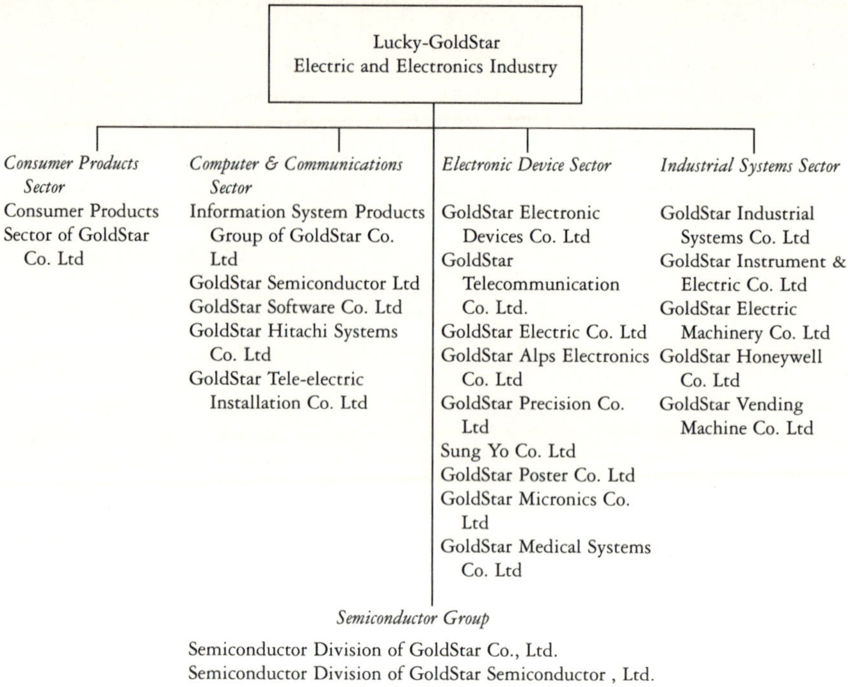

Figure 8.2 1987 reorganisation of GoldStar Consumer and Industrial Electronics
Source: Adapted from *GoldStar Company Report* (1988, Seoul), pp. 6–7; Exhibit 4–4, at p. 62 *The Chaebols (Steers, Shin & Ungson 1989).*

become a major production centre. It was the first major manufacturing facility built in the United States by a Korean company. A second facility was opened in Worms, West Germany, in 1987 with an annual capacity of 300,000 televisions and 400,000 VCRs.

The electronics section of Lucky-GoldStar underwent a large change in organisation in 1986–7. During this period, as Figure 8.2 illustrates, the major divisions of the GoldStar (Electronics) Co. were reorganised into five groups in order to take advantage of the dynamic changes both in technology and the marketplace in electronics with more efficiency.

As a result, GoldStar achieved rapid growth in the sectors of consumer products, computers and communications, electronic devices, industrial systems and semiconductors. GoldStar (Electronics Co.) was ranked 279th in the world in terms of total sales in the 1993 Fortune International 500. Among electronics and electrical equipment corporations, GoldStar was 34th largest in the world in 1993 with total sales of US$ 5,367 million.

8.2 A review of the findings of the survey and the interviews

8.2.1 Profile of Korean consumer electronics firms in the EU

(i) Medium-sized Korean consumer electronics firms predominated in the EU.

A total of 16 companies answered the question on the number of employees, as Table 8.4 shows. The answer shows that medium-sized Korean consumer electronics firms predominate in the EU, which is also true of the Japanese manufacturing firms in Europe. In fact, as of the end of 1994, the average number of employees of Korean consumer electronics firms in the EU was 260, compared to the average number of 294 for Japanese manufacturing firms in Europe as of the end of 1991.

By capital amount, as of the end of 1994, Korean manufacturing firms in the EU were also medium-sized, with average capital of US$ 10.37 million, as Table 8.5 shows. The average capital of Korean consumer electronics firms in the EU was much smaller than that of Japanese equivalents, whose the average capital was US$ 17.8 million as of the end of 1991. Four out of 13 companies (31 per cent) had capital in the range of between US$ 1 million and US$ 5 million, with another four holding capital of between US$ 5 million and US$ 10 million, while no company had more than US$ 100 million.

The average number of employees in Korean consumer electronics firms in Germany was 539, a larger number than in any other state, as Table 8.6

Table 8.4 Size of Korean consumer electronics firms in the EU (employee numbers)

	Less than 49 employees	*50–99 employees*	*100–199 employees*	*200–299 employees*	*300–499 employees*	*500–999 employees*	*Average*
No. of companies	3	3	2	2	5	1	260

Note: 16 firms responded.

Table 8.5 Size of Korean consumer electronics firms in the EU (capital)

	Less than 1 million US$	*1.01–4.99 million US$*	*5–9.99 million US$*	*10–49.99 million US$*	*50–99.99 million US$*	*More than 100 million US$*	*Average (US$ million)*
No. of companies	1	4	4	3	1	0	10.37

Note: 13 firms responded.

shows. The UK accounted for the largest number of FDI (seven cases), 44 per cent of the total FDI in the EU.

The average number of Korean employees at a manager-level or higher position was seven per company, while that of lower positions was only one, as Table 8.7 shows. The average number of local employees of manager-level or higher position was 20, while that of lower position than managers was 231. As the figures show, most of higher positions were taken by Koreans while only 8 per cent of all locally hired employees took higher positions. According to the results of survey and interviews, the main job of Korean personnel was to support performance of local staff by playing a bridging role between parent companies and local staff. However, Korean personnel usually make important decisions on business in the country where they are doing local production. This is no different from Japanese or German companies operating in foreign countries. British and Swedish firms also fit this pattern, but to a lesser degree.

The actual local business activities, including personnel administration, accounting, general affairs, manufacturing and public relations, are normally assigned to local staff. In particular, 88 per cent of Korean consumer electronics companies assigned the job of personnel administration to a locally employed manager. Only one company (Saronix Kony) assigned R&D jobs to local staff. This company was, in fact, totally run by local staff. In addition, among these 16 companies, only one company (Lucky-GoldStar) had a large-scale R&D centre in the UK.

Table 8.6 Number of employees of Korean consumer electronics firms in the EU by country (responding firms only)

	No. of firms responded	*Average no. of employees per firm*
UK	7	166
Rep. of Ireland	2	264
France	1	40
Germany	3	539
Italy	1	175
Spain	1	215
Portugal	1	389
Total/average	16	260

Table 8.7 Korean employees and local employees by position

	Average no. of manager-level staff and higher	*Average no. of staff (lower position than managers)*
Korean staff	7	1
local staff	20	231

(ii) Principal competitors of Korean consumer electronics firms in the EU were Japanese firms. Korean consumer electronics firms in the EU thought Korean firms' brand names or marketing techniques and accumulated human capital resources were the biggest competitive advantages over competitors (Japanese, American and European companies) in the EU.

With regard to Korean firms' principal competitors in the EU, 14 firms replied (plural answers allowed). Japanese companies were regarded as the

Table 8.8 Divisions under the control of a locally recruited head

Divisions under the control of locally recruited head	*No. of companies*
Personnel administration	14 (88%)
General affairs	13 (81%)
Public relations	11 (69%)
Marketing	6 (38%)
Manufacturing	12 (75%)
Accounting	13 (81%)
Financing	5 (31%)
R&D	1 (6%)

strongest competitors in the EU by eight Korean consumer electronics companies, accounting for 57 per cent of the 14 firms with valid replies. Two companies in particular (GoldStar Iberna Italy SPA,[10] and Samsung Corning Deutschland GmbH) answered that there were no competitors to their companies in the EU.

In reply to a question regarding competitive advantage in the EU, 16 companies supplied answers, as Table 8.9 shows. The findings were that brand-name or marketing techniques and accumulated human capital resources were considered the most significant competitive advantages over competitors in the EU. In seven areas given as the list of examples in the questionnaire, the companies did not show big differences in their answers on their competitive edge over competitors in the EU. The average value of all companies was 2.6. According to the scale given in the questionnaire (1 = no advantage, 2 = somewhat important, 3 = important, 4 = very important, 5 = most important, and 0 = don't know), this meant that Korean consumer electronics companies operating in the EU thought they had *some* competitive advantages over the competitors such as Japanese, American and British firms in the EU.

(iii) Most Korean consumer electronics companies manufacture end-user products in the EU.

Regarding the products of Korean consumer electronics companies, 69 per cent – nine companies out of 16 – replied that they manufactured end-user

Table 8.9 Competitive advantage of Korean consumer electronics firms in the EU over competitors in the EU

	a	*b*	*c*	*d*	*e*	*f*	*g*	*h*	*i*	*j*	*k*	*l*	*m*	*n*	*o*	*p*	*A**
A	3	3	3	0	5	1	2	4	1	3	4	3	3	3	2	4	2.9
B	3	2	2	3	2	1	3	2	1	3	3	3	4	2	5	0	2.6
C	2	3	2	5	3	4	2	3	1	1	3	2	2	3	3	0	2.6
D	1	2	2	4	3	1	2	3	1	3	3	2	4	3	4	0	2.6
E	2	1	2	3	3	1	2	3	1	1	3	2	4	2	3	0	2.2
F	2	1	2	4	3	3	1	2	1	4	4	2	4	2	3	0	2.5
G	3	2	3	5	5	1	2	1	4	2	2	1	1	3	5	4	2.8

Notes: *a*: Kolon Imperial Graphic PLC. *b*: SKM Europe Ltd. *c*: Inkel (UK) Ltd. *d*: GoldStar Electric (UK) Ltd. *e*: Daewoo Electronics UK Ltd. *f*: Trigem Computers (UK) Ltd. *g*: Samsung Electronics UK Ltd. *h*: Haitai Europe S.A. *i*: Saehan Media Ireland Ltd. *j*: SaRonix Kony. *k*: GoldStar Europe GMBH. *l*: Samsung Elektronische Bauelemente. *m*: Samsung Corning Deutschland GMBH. *n*: Samsung Portugal Produtos Electronecanicos S.A. *o*: GoldStar Iberna Italy SPA. *p*: Samsung Electronica Espanola S.A. *A**: average point of answers. '0', which means 'don't know', was not included in the calculation of averages.

A: marketing technique or brand names. B: appropriate technology. C: exclusive access to indispensable inputs. D: experience of major technical improvement on production processes or new products. E: new technology in developing new material or resources saving innovations. F: possession of capacity to innovate (e.g. because of sufficient investment in R&D). G: accumulated human capital resulting in low management cost.

products. Only four companies (about 6 per cent) produced parts and components. Two companies (Saehan Media Ireland Ltd *and* Samsung Electronica Espanola, S.A.) particularly focused on producing both semi-finished goods and end-user products. This meant that 69 per cent of Korean consumer electronics companies in the EU simply assembled parts and components for end-user products in the EU.

(iv) 67 per cent of Korean consumer electronics firms in the EU obtained more than 50 per cent of the necessary capital for setting-up the firms in the EU from their parent companies.

Regarding the source of capital for setting-up firms in the EU, 15 companies supplied answers. Ten of these companies (67 per cent) got more than 50 per cent of the necessary capital from their parent company. In particular, among these ten companies, eight got 100 per cent capital for setting-up their firms in the EU from their parent companies. Six companies (40 per cent) used the local financing market as the secondary source of capital for setting-up their firms in the EU. This is exactly coincident with the

Table 8.10 Sales characteristics of Korean consumer electronics firms in the EU

	a	*b*	*c*	*d*	*e*	*f*	*g*	*h*	*i*	*j*	*k*	*l*	*m*	*n*	*o*	*p*
A	23: 77	100: 0	20: 80	50: 50	25: 75	40: 60	30: 70	70: 30	0: 100	5: 95	40: 60	10: 90	85: 15	0: 100	20: 80	20: 80
B	No	–	No	No	No	No	—	No	No	No	No	No	No	Yes	No	No

Notes: A: ratio between local sale and exports. B: existence of compulsory minimum exporting percentage in the country where the company is operating.
Column headings as Table 8.9.

theory on the financing of foreign direct investment. According to this theory, 'after an initial start, affiliates abroad must normally rely on internally generated funds and local financing through borrowing and supplier credits' (Gilman 1981: 65). Only one company used Korean banks as the secondary source of capital requested for setting-up their firms. Two companies used host government grants as their third source of necessary capital. One company (Samsung Portugal Produtos Electronecanicos S.A.) used loans raised in international financial markets as its fourth source of necessary capital.

(v) 75 per cent of Korean consumer electronics firms (12 out of 16 companies) in the EU exported more than 50 per cent of their products in the EU.

To the question about the ratio of local sales to exports, a total of 16 companies replied. Of these, 75 per cent (12 companies) exported more than 50 per cent of their products, as illustrated in Table 8.9. Sixty-three per cent (10 companies) exported more than 70 per cent of their products. Only 19 per cent (three companies) sold more products in local markets than are exported to foreign markets. One company (GoldStar Electric UK Ltd) sold its products evenly (50 per cent each way) in local and foreign markets.

Fourteen companies replied to the question concerning the existence of a compulsory minimum exporting percentage in the country where they were operating. Among these companies, only one (Samsung Portugal Produtos Electronecanicos S.A.) stated that there was a compulsory minimum exporting percentage (21 per cent in 1993). The remaining 13 companies stated that there were no compulsory minimum exporting percentages in countries where they were operating. This suggested that the companies entered a specific EU member state in order to take advantage of favourable production conditions and to facilitate export to third countries, and not necessarily to dominate the local market.

(vi) As of the end of 1994, 69 per cent of Korean consumer electronics firms reached the break-even point or entered into the black. In reply to the major reasons for the losses of Korean consumer electronics firms in the EU, the companies pointed out the short operation period.

A total of 16 companies supplied answers relating to their operational results. Only one replied that its European business had been consistently profitable. Four companies answered that their European businesses had been consistently loss-making. However, 69 per cent of firms (11 companies) found that their European businesses were coming out of the red and reaching the break-even point or becoming profitable.

With regard to the reasons for operational losses, 15 companies replied, as Table 8.11 shows, that the major reason was that their business was still in its infancy. In fact, only one company (Samsung Electronics UK Ltd) started its business in 1984 while the others started in the late 1980s and the beginning

Table 8.11 Reasons for losses made by Korean consumer electronics firms in the EU

	a	*b*	*c*	*d*	*e*	*f*	*g*	*h*	*i*	*j*	*k*	*l*	*m*	*n*	*o*	*p*	A
1	4	3	3	5	4	4	2	4	5	4	2	5	2	–	5	4	3.7
2	1	2	1	1	4	1	3	5	3	2	5	5	4	–	1	1	2.6
3	1	1	3	1	3	3	1	1	1	3	2	1	1	–	1	1	1.6
4	1	2	1	4	2	1	1	2	5	2	2	1	0	–	1	1	1.9
5	2	2	2	5	3	1	–	2	3	2	2	1	2	–	3	5	2.5
6	1	4	1	1	3	2	3	3	1	3	3	1	1	–	1	5	2.2
7	2	4	1	–	2	3	1	4	1	3	4	1	2	–	2	2	2.3

Notes: '0', which means 'don't know', was not included in the calculation of average.
1: The business is still in its infancy. *2*: Higher local labour costs. *3*: Small market. *4*: Reduced market share due to intensified competition with other Korean firms in Europe. *5*: Reduced market share due to intensified competition with European companies. *6*: Deteriorating competitiveness. *7*: higher financial costs, such as interest cost.
Column headings as Table 8.9.

of 1990s, as illustrated in Table 8.12. The second major reason for the losses of Korean firms in the EU businesses was high labour costs. The third was the reduced market share which had resulted from intensified competition with European firms.

In particular, the answers given by the four firms which had been consistently loss-making were quite similar to the answers of all 15 responding Korean consumer electronics firms in the EU. Those four firms answered that both high labour costs in the EU and the short history of the business there were the major reasons for operational losses. The Reddaway report (Reddaway *et al.* 1968) on British investment in the 1960s also concluded

Table 8.12 Starting date of business of 16 responding Korean consumer electronics firms (including parts and components) in the EU

Name of Company	*Country of operation*	*Operation started*
Samsung Electronics UK Ltd	UK	7 March 1984
GoldStar Europe GmbH	Germany	4 September 1986
Saehan Media Ireland Ltd	Rep. Of Ireland	21 April 1987
GoldStar Electric UK Ltd	UK	13 October 1988
Daewoo Electronic UK Ltd	UK	5 November 1988
Samsung Electronica Espanola S.A.	Spain	17 January 1989
Trigem Computers UK Ltd	UK	8 September 1989
Haitai Europe	France	27 September 1989
GoldStar Iberna Italy SPA	Italy	25 January 1990
Samsung Portugal Produtos Electronecanicos (parts manufacturer)	Portugal	22 May 1990
SaRonix Kony (parts manufacturer)	Rep. Of Ireland	12 June 1990
Inkel UK Ltd	UK	10 July 1990
SKM Europe UK Ltd	UK	30 May 1991
Kolon Imperial Graphics UK PLC	UK	13 September 1991
Samsung Elektronische Bauelemente (parts manufacturer)	Germany	September 1992
Samsung Corning Deutschland GmbH (parts manufacturer)	Germany	1 March 1994

that 10 years was the average time it took to break even with a 'green field' foreign affiliate. Deteriorating competitiveness was the third reason for operational losses in the EU.

(vii) 50 per cent of the responding 16 Korean consumer electronics companies out of the total of 26[11] *thought that the integration of the EU market had not affected them.*

In reply to the question about the effects on Korean firms in the EU of the integration of the EU market, again 16 companies answered. The results showed that the Korean consumer electronics firms thought that the integration of the EU market would have both positive and negative effects. As Table 8.13 illustrates, the main four potential effects found by Korean consumer electronics companies were both positive and negative: (1) unified certification system (CE mark) would be introduced (positive effect), (2) sentiment would be oriented towards trade protectionism (negative effect), (3) competition with Japanese, European and American companies in the EU would intensify (negative effect) and (4) unification of safety, health and environmental standards would make intra-distribution easier and customs formalities would be simplified (positive effect).

However, Korean consumer electronics firms in the EU believed that there would be more positive effects overall from the integration of the EU market. With regard to the actual effects which the integration process of the EU market already had on the Korean consumer electronics firms, 16 companies replied (some duplicated). 50 per cent (eight companies) replied that the integration of the EU market had not yet affected their business in the EU, while 38 per cent (six companies) said that it had already had a positive effect on their business in the EU. Only 19 per cent (three companies) said that the integration of the EU market had a negative effect on their business in the EU. The results here were contradictory to their answers to the question on their motives of advancing into the EU.

Table 8.13 Potential effects of the integration of the EU market on Korean consumer electronics firms in the EU

	a	*b*	*c*	*d*	*e*	*f*	*g*	*h*	*i*	*j*	*k*	*l*	*m*	*n*	*o*	*p*	A
P																	
3	4	2	4	2	4	2	1	2	1	4	2	2	2	3	2	5	2.6
4	3	3	4	4	3	3	2	2	3	4	3	2	3	4	3	2	3
7	3	4	3	3	4	4	3	2	1	4	4	1	2	3	2	1	2.7
8	2	2	2	1	3	2	1	1	1	4	2	3	1	3	3	1	2
10	2	3	2	1	4	5	2	3	1	1	4	–	2	3	1	1	2.3
11	4	3	3	5	3	5	2	2	1	3	4	3	3	3	3	3	3.1
12	4	4	3	3	3	5	2	4	3	4	5	1	1	3	3	5	3.3
13	2	2	3	1	4	3	2	2	3	1	2	1	1	1	2	1	1.9
15	2	4	3	5	–	4	2	4	4	4	4	3	2	3	4	4	3.3
16	4	4	3	5	–	4	2	5	1	4	4	–	4	3	4	5	3.7
N																	
1	3	2	3	5	4	4	2	5	5	4	4	3	4	3	4	3	3.6
2	2	2	3	4	3	3	2	5	1	2	4	3	4	3	3	3	2.9
5	4	2	3	5	4	4	3	5	2	5	4	1	3	2	3	4	3.4
6	4	1	2	4	3	4	1	2	3	2	3	3	3	2	2	1	2.5
9	3	2	3	5	4	4	3	1	4	3	4	3	5	2	3	2	3.2

Notes: P: positive effects. N: negative effects. 1: Sentiment will be oriented towards trade protectionism. 2: There will be a move towards demanding reciprocity. 3: Generally, there will be a move towards expanded business opportunities. 4: EU firms will be vitalised, 5: Competition with Japanese, European and American companies in the EU will intensify, 6: Competition with other Korean firms will intensify. 7: Administrative procedures will be simplified. 8: Trade barriers will be removed. 9: There will be a move towards shutting Korean firms out of the EU market. 10: There will be a chance to enter the public procurement market and participate in public works. 11: There will be changes in patterns of physical distribution in the EU. 12: Unification of safety, health and environmental standards will make intra-distribution easier. 13: Import quota system applied to Korea will be removed. 14: Unification of taxation system may remove benefits from tax incentives. 15: Customs formalities will be simplified. 16: Unified certification system (CE mark) will be introduced. Column headings as Table 8.9.

Table 8.14 Actual positive and negative effects of the integration of EU market on the business of Korean consumer electronics firms in the EU

	a	*b*	*c*	*d*	*e*	*f*	*g*	*h*	*i*	*j*	*k*	*l*	*m*	*n*	*o*	*p*	*A*
P			√	√				√		√		√		√			
1	2	–	4	5	5	5	–	5	4	4	–	5	4	4	3	1	3.9
2	3	–	4	5	3	1	–	4	–	4	–	5	1	4	3	1	3.2
3	1	–	3	–	–	5	–	1	–	4	–	4	1	1	1	1	2.2
4	1	–	3	4	–	1	–	1	–	2	–	4	1	1	1	1	1.8
U	√	√			√		√		√				√		√	√	
	a	*b*	*c*	*d*	*e*	*f*	*g*	*h*	*i*	*j*	*k*	*l*	*m*	*n*	*o*	*p*	*A*
N				√		√					√						
5	1	–	3	0	–	0	–	1	0	2	3	0	1	–	0	1	1
6	1	–	3	4	–	0	–	2	0	3	2	0	1	–	2	1	1.6
7	1	–	3	5	–	0	–	4	0	3	4	0	1	–	2	1	2

Notes: '0', which means 'don't know' is not included in the calculation of average. P: Positive effect on the business in the EU. U: No effect on the business in the EU. N: Negative effect on the business in the EU. *1:* Easier free distribution in the EU. *2:* Reduction of distribution costs in the EU. *3:* Will become possible to enter into the public procurement market. *4:* Reduction of production costs. *5:* Easier making profits in other countries due to delays in legislating EU directives in host countries. *6:* Negative effects due to inadequate understanding of procedures to apply for the basic certification regime. *7:* Weakening competitive ability due to the favour which European enterprises get from the integration of EU market. Column headings as Table 8.9.

With regard to the actual positive effect on their business in the EU, 13 companies replied, as illustrated in Table 8.14. Of the actual positive effects of the European market integration on Korean enterprises, the two most important items were (1) to make free distribution of products easier in the EU and (2) to reduce the distribution costs in the EU. Concerning the actual negative effect on their business in the EU, 13 Korean consumer electronics firms supplied answers. According to them, the main negative effect was the relative reduction of their competitive ability in the EU market due to the favour which European firms received from the integration of the EU market.

8.2.2 *Corporate strategies of Korean consumer electronics firms in the EU*

(i) The major motive of advancing into the EU was to follow the so-called 'globalization' strategy and the major reason for selecting a specific EU member country was how favourable host country grants (subsidies) were.

Asked about their motives and aims in making FDI in the EU, 16 out of 26 companies submitted replies. As illustrated in Table 8.15, the top four

Table 8.15 Top five motives and aims of Korean consumer electronics firms in the EU

	a	*b*	*c*	*d*	*e*	*f*	*g*	*h*	*i*	*j*	*k*	*l*	*m*	*n*	*o*	*p*	A
1	4	2	3	5	5	5	2	4	4	5	4	5	5	4	5	4	4.1
2	4	4	5	5	5	2	1	5	5	1	3	5	4	3	3	5	3.8
3	3	3	3	5	4	5	1	5	4	1	5	5	5	4	3	4	3.8
4	4	3	3	3	4	3	1	5	4	3	4	5	3	4	3	5	3.6
5	2	3	3	5	4	3	2	5	4	–	–	0	1	1	2	3	2.9

Notes: '0', which means 'Don't know', is not included in the calculation of averages.
1: To carry out part of globalization. *2:* To avoid discriminatory restrictions such as tariffs and quotas on Korean imports by the EU member states. *3:* To protect the existing market in the EU and retain or increase market shares from the expansion of economic activity resulting from the EU's market integration. *4:* Concern about a rise in protectionism as a result of EU market integration. *5:* To avoid anti-dumping decisions against Korean made components. Column headings as Table 8.9.

motives or purposes for advancement were: (1) to carry out part of a globalisation strategy; (2) to avoid discriminatory restrictions, such as tariffs and quotas on Korean imports by the EU member states; (3) to protect the existing market in the EU and retain or increase market shares from the expansion of economic activity resulting from the EU's market integration and (4) to act on concerns about a rise in protectionism as a result of the EU's market integration, and to avoid anti-dumping decisions against Korean made components.

These survey results showed that Korean consumer electronics firms advanced into the EU primarily to become multinational companies operating all over the world. In fact, most Korean *chaebols* have focused on domestic market activities under the special protection of the Korean government and on several foreign markets, such as the American or Japanese market. However, the interviews with managing directors of Korean consumer electronics firms in the EU revealed that the Korean firms started feeling the necessity for a so-called globalisation strategy because of their growing concern about protectionism arising in the EU as a result of its market integration. In addition, in the questionnaire, the companies also implied the same thing.

In reply to the question about reasons for selecting specific European countries in which to invest, a total of 16 firms supplied answers, as Table 8.16 shows. The first major reason for locating in a specific country in the EU was the offer of host country government grants (subsidies) at more favourable conditions than the other EU countries.

The average value attached to this, according to the scale in the questionnaire (1 = of no importance, 2 = somewhat important, 3 = important, 4 = very important, 5 = most important, 0 = don't know), was 3.4. This meant the offer of government grants (subsidies) at favourable conditions was quite

Table 8.16 Reasons for selecting specific European countries to advance into

	a	*b*	*c*	*d*	*e*	*f*	*g*	*h*	*i*	*j*	*k*	*l*	*m*	*n*	*o*	*p*	A
1	1	3	3	5	5	4	1	4	1	1	3	1	4	1	3	1	2.6
2	1	1	2	3	3	1	2	1	1	1	4	3	2	1	3	1	1.9
3	4	2	4	5	5	3	2	1	1	3	3	1	1	1	1	1	2.4
4	4	2	3	5	4	4	2	1	3	3	2	0	1	3	2	4	2.7
5	2	1	3	5	2	3	2	1	4	3	3	1	1	1	1	2	2.2
6	4	1	4	4	2	3	1	5	4	4	3	3	4	1	1	2	2.9
7	3	1	4	5	3	3	2	3	3	4	4	4	2	1	1	4	2.9
8	3	2	3	3	3	2	2	2	1	2	4	3	2	1	2	4	2.4
9	1	1	1	1	2	3	1	1	1	1	2	0	1	1	1	1	1.3
10	1	1	1	1	2	3	2	1	1	2	2	0	1	1	1	4	1.6
11	3	2	3	5	3	1	1	4	3	3	2	0	1	3	2	2	2.5
12	3	–	3	4	3	1	2	4	2	3	3	4	2	1	1	4	2.7
13	1	1	3	1	2	1	1	5	1	2	2	1	1	1	1	1	1.6
14	1	3	3	3	2	2	2	4	1	1	3	4	4	1	3	4	2.6
15	2	1	3	1	2	1	2	1	2	0	3	1	1	2	3	4	1.9
16	2	1	3	4	3	1	1	1	1	1	2	0	1	1	1	2	1.7
17	2	1	1	1	3	1	1	1	1	2	4	1	1	3	3	3	1.8
18	2	1	3	3	3	1	1	2	1	3	4	1	1	2	2	4	2.1
19	2	1	3	3	2	1	2	5	4	4	3	3	5	1	2	1	2.6
20	2	1	4	4	4	1	2	5	5	5	3	3	5	4	5	1	3.4
21	1	1	3	3	2	1	2	3	5	5	2	0	1	4	3	1	2.5

Notes: '0', which means 'don't know', was not included in the calculation of averages. *1:* Larger market size of the host country. *2:* Availability of advanced technologies. *3:* Availability of English-speaking people. *4:* Cheaper labour costs. *5:* Cheaper capital costs. *6:* Favourable investment environment in terms of infrastructure, government efficiency or political stability. *7:* Better quality of labour. *8:* Availability and favourable prices of necessary raw materials and/or parts and components. *9:* Operation of many Korean firms in the country. *10:* Operation of foreign competitors, especially competitors from Japanese and Asian NICs. *11:* Easy export to third countries. *12:* Favourable geographical conditions for physical distribution. *13:* Existence of stronger tariff and non-tariff barriers to trade in the host country. *14:* More important position of the host country in the world market (market share, competitiveness, growth potential). *15:* Prevalent pro-Korean sentiment in the host country. *16:* Convenience in the education of children. *17:* Economic development stage of the host country suitable to the lower end of technology and capital equipment which are becoming outdated in the home country. *18:* Macro-economic policy of the host country which promotes economic stability and growth potential. *19:* More favourable tax incentives. *20:* Offers of government grants at more favourable conditions. *21:* Easy access to EU restructuring funds. Column headings as Table 8.9.

an important factor which affected their selection of FDI location in the EU. As the second major reason, the companies pointed to a favourable investment environment in terms of infrastructure, government efficiency or political stability, and good quality of labour.

(ii) Specific measures which Korean consumer electronics firms were taking to cope with the integration of the EU market, included 'Europeanising' their operations by increasing the local content of parts and raw materials, and placing locally employed people on higher rungs of the corporate ladder, and by making a contribution to the local community.

With regard to specific measures taken by Korean consumer electronics firms to cope with the integration of EU market, 16 companies replied (some duplicated). As Table 8.17 shows, the measure which the largest number of companies (13 companies) took was to 'Europeanise' operations by increasing the local content of parts and raw materials, by placing local employees higher on the corporate ladder, and by contributing to the local community.

In addition, eight companies were setting up or considering setting up a controlled company in the EU member states which would be independently responsible for production, sales, fundraising and R&D activities. As the third popular measure, seven companies established or considered establishing design centre functions and facilities in the EU in order to be more responsive to European design preferences.

Table 8.17 Specific measures which Korean consumer electronics firms are taking to cope with the integration of EU market

	a	*b*	*c*	*d*	*e*	*f*	*g*	*h*	*i*	*j*	*k*	*l*	*m*	*n*	*o*	*p*	N
1		√		√				√		√	√	√		√	√		8
2			√	√				√		√	√			√	√		7
3	√		√	√	√	√	√	√	√		√	√		√	√	√	13
4			√	√									√		√	√	5
5			√	√	√		√										4
6	√	√						√		√			√		√		6
7	√			√										√	√	√	5
8											√						1
9				√		√		√		√					√		5

Notes: N: Number of companies taking the measure. 1: Setting up or considering setting up a controlling company in the EU member states independently responsible for production, sales, fund raising and R&D activities. 2: Establishing design centre functions and facilities in the EU. 3: To make 'Europeanised' corporations by increasing the local content of parts, placing locally employed people on the higher rungs of the corporate ladder and making a contribution to the local communities. 4: Increasing production shares in the EU by increasing productive capacity and setting up production bases in the EU. 5: Bring in parts suppliers and requesting Korean parts manufacturers to advance into the EU in an attempt to increase local content of parts and components. 6: Considering close co-operation with European firms in production and sales and M&A with European companies, when the occasion demands it. 7: Sending competent persons to the EU as resident representatives. 8: Considering shifting to new production base in other country which offers more favourable conditions in labour cost and tax incentives. 9: Considering production of products adapted to the unified certification system (CE mark) Column headings as Table 8.9.

(iii) *As a preferred method of expansion, 67 per cent of Korean consumer electronics firms in the EU chose 'expansion of existing business site'.*

With regard to the preferred method of expansion in the EU, 15 companies answered (some duplicated). As Table 8.18 shows, 67 per cent (10 companies) said that their preferred method of expansion in the EU was through the 'expansion of the existing business sites (simple purchase of land or buildings)'. A 'joint venture' was the second preferred method of expansion.

Table 8.18 Preferred method of expansion of Korean consumer electronics firms in the EU

a	*b*	*c*	*d*	*e*	*f*	*g*
10	2	1	3	0	1	1

Notes: *a*: Expansion of existing business site. *b*: Acquisition of competitor. *c*: Acquisition in new area of operations. *d*: Joint venture. *e*: Franchising/licensing operation. *f*: Contracting with local firm(s) to provide management services/representation. *g*: No answer.

(iv) *Korean consumer electronics companies answered that the host country government was the most relevant body with which they discussed their business in the EU. As the most helpful intermediaries in explaining and communicating the interests of Korean consumer electronics firms to the relevant bodies, 67 per cent of them chose host country government departments ministries.*

Concerning the relevant bodies with which Korean consumer electronics firms in the EU discussed their business, a total of 16 companies supplied answers. Of these, 50 per cent said the host country government was their preferred counterpart, as illustrated in Table 8.19. At the same time, the companies had some problems in finding the right bodies with whom they could discuss their business in the EU. Four companies said that they were not certain which bodies were responsible for which policies.

In reply to the question on which bodies were the most helpful intermediaries in explaining and communicating the companies to relevant bodies, 67 per cent (10 companies) said that the host country government/ministry was the most helpful intermediary in explaining and communicating their interests to the relevant bodies, as illustrated in Table 8.20.

(v) *The expected pay-off period for investment for 56 per cent of companies (nine companies) was more than five years.*

All 16 companies gave the answer to the question asking about the expected pay-off period for their investment. Nine replied that they expected the initial investment to be recovered in more than five years later. Seven

Table 8.19 Relevant organisations with which Korean consumer electronics firms discuss business-related issues (plural answers allowed)

	a	*b*	*c*	*d*	*e*	*f*	*g*	*h*	*i*	*j*	*k*	*l*	*m*	*n*	*o*	*p*	N
1					√			√			√						3
2																	0
3									√						√		2
4			√	√				√	√			√		√	√	√	8
5	√			√									√				3
6											√		√				2
7											√	√					2
8		√				√	√				√						4

Notes: 1: EU Commission. 2: European Parliament. 3: Council of Ministers/member government. 4: Host country government. 5: Korean government. 6: KOTRA (Korea Trade Promotion Corporation). 7: Korean trade association. 8: Not yet certain which bodies are responsible for which policies. N: Number of companies selecting each example.
Column headings as Table 8.9.

Table 8.20 The most helpful intermediaries in explaining policy and communicating between Korean consumer electronics firms in the EU and the relevant bodies (plural answers allowed)

	a	*b*	*c*	*d*	*e*	*f*	*g*	*h*	*i*	*j*	*k*	*l*	*m*	*n*	*o*	*p*	N
1		√	√	√	√		–	√	√	√	√			√		√	10
2							–										0
3						√	–										1
4							–	√							√		2
5							–										0
6							–								√		1
7							–					√	√			√	3
8	√						–			√							2

Notes: 1: Host country government department/ministry. 2: Korean government/embassy. 3: KOTRA (Korea Trade Promotion Corporation). 4: Consultant/lobbyist. 5: Korean trade/industry association. 6: European Union trade/industry association. 7: Direct approach sufficient; intermediary unnecessary. 8: None of the above. N: Number of companies selecting each example.
Column headings as Table 8.9.

companies thought that the expected pay-off period for investment would between three and five years. No company replied that the initial investment would be recovered within three years.

8.2.3 *Efforts of Korean consumer electronics firms to be more integrated into the host country's economy and society*

(i) *All 16 Korean consumer electronics firms in the EU favoured the employment of host country nationals for managerial positions.*

A total of 16 firms answered the question regarding the appointment of locally recruited personnel as managers, and all 16 companies gave an affirmative reply. They said the major reasons for their preference for the appointment of locally recruited personnel as managers were: (1) to save operational costs by hiring local personnel whose salary level was relatively lower than Korean expatriates on a short-term basis and (2) to achieve localisation of their companies on a long-term basis.

With respect to the highest position of locally recruited people, 50 per cent of the companies (eight) said these people were appointed as 'managers'. This was very different from the Japanese case. According to the result of a survey conducted by JETRO in 1992, 28.5 per cent of the Japanese firms, the largest percentage, said that these people were appointed as 'chief executive officers'. Four of the Korean firms (25 per cent) answered that directorship was the highest position held by locally recruited people. A further 19 per cent (three companies) said they appointed a local chief executive officer. Only one company said that supervisorship was the highest position for locally recruited personnel.

(ii) *94 per cent of the parent companies of Korean consumer electronics firms gave authority over managerial appointments to their European affiliates.*

In reply to the question of whether adequate authority was delegated by the parent company or not, again a total of 16 companies replied. Of these, 15 replied in the affirmative, while only one company (Daewoo Electronics UK Ltd) did not. According to the answers given by the 15 companies which felt they had adequate authority, they were almost completely independent from the parent companies when it came to making new products, as Table 8.21 illustrates. In appointing or dismissing staff, companies gave an average value of '2.3'. This meant, according to the scale given in the questionnaire (1 = small powers delegated, 2 = quite large powers delegated, 3 = all powers delegated, 0 = don't know), that the companies overall were fairly independent from parent companies.

Table 8.21 Powers delegated by parent company

	a	*b*	*c*	*d*	*e*	*f*	*g*	*h*	*i*	*j*	*k*	*l*	*m*	*n*	*o*	*p*	A
1	1	1	1	2	–	√	1	√	0	3	1	√	2	2	1	1	1.3
2	2	3	1	3	–	3	1	3	1	3	2	2	3	3	2	3	2.3
3	1	1	1	2	–	√	1	1	0	3	1	2	2	2	2	2	1.5
4	1	1	2	2	–	√	2	2	2	3	2	3	2	2	2	2	2
5	3	3	3	3	–	√	2	2	3	3	2	3	2	2	2	3	2.6
6	2	3	3	2	–	2	1	3	3	3	1	3	2	1	2	1	2.1
7	1	1	1	2	–	√	2	1	3	2	1	2	2	2	2	2	1.7

Note: 1: To change capital stock. 2: To appoint or dismiss officers. 3: To decide how to use profits made from business. 4: To make investment in plant and equipment. 5: To change manufacturing process. 6: To market new products. 7: To determine loans and investments programmes. Do not know. This is not included in the calculation of the average point of answers.
Column headings as Table 8.9.

(iii) 64 per cent of Korean consumer electronics firms in the EU participated in local cultural and social activities in order to localise their companies, to improve their image and strengthen the relations with universities, which would improve their R&D capability.

In reply to the question of whether Korean consumer electronics firms participated in cultural and social activities in the EU, 14 companies replied. Of these, 64 per cent (nine companies) said that they did participate in cultural and social activities locally; in particular, all three GoldStar companies and four of the five Samsung companies said they did this. The major reason given by GoldStar for their participation in these activities was to strengthen relations with higher educational institutes like universities, which would result in improved of R&D capability. In the case of the Samsung companies, the major reason for the participation in these activities was to improve the company image in the countries in which they were doing business. Besides these reasons, many companies said that they undertook such activities in order to achieve localisation of their companies.

(iv) Concerning the existence of compulsory minimum local content ratio in the country where they were operating, 62 per cent of companies reported the existence of such ratios. With regard to source of material supply, 69 per cent of companies said that parent companies were major suppliers. Relating to local content ratios of parts, components and materials used by the firms in the EU, 69 per cent of companies said the ratio was increasing, primarily due to on-time delivery of local suppliers.

With regard to the existence of compulsory minimum local content ratio in the country where they were operating, 13 companies replied. Of these, 62

Table 8.22 Major reasons for the increase of local content ratio of Korean consumer electronics firms in the EU

	a	*b*	*c*	*d*	*e*	*f*	*g*	*h*	*i*	*j*	*k*	*A*
1	1	1	3	2	1	1	4	2	1	1	3	1.8
2	2	1	4	2	2	4	5	3	4	1	3	2.8
3	1	4	2	1	5	0	4	1	3	1	5	2.5
4	1	4	3	2	1	3	4	2	1	1	2	2.2
5	1	1	4	2	1	–	4	1	1	1	4	2
6	1	1	1	1	1	0	2	1	1	1	1	1
7	1	1	1	1	1	1	2	1	5	1	1	1.5
8	1	1	2	1	3	1	2	1	1	1	1	1.4
9	1	1	1	1	5	5	2	4	1	1	1	2.1

Notes: *a*: SKM Europe Ltd. *b*: Saehan Media Ireland Ltd. *c*: SaRonix Kony Ireland Ltd. *d*: GoldStar Iberna Italy SPA. *e*: Inkel (UK) Ltd. *f*: GoldStar Electric (UK) Ltd. *g*: Daewoo Electronics UK Ltd. *h*: Samsung Portugal Produtos Electronecanicos SA. *i*: Trigem Computers (UK) Ltd. *j*: Samsung Electronics UK Ltd. *k*: Samsung Electronica Espanola S.A.
1: Quality improvement of locally produced parts, components and materials. *2:* On-time delivery of local suppliers. *3:* Price reduction of locally produced parts, components and materials. *4:* Required supplies can be procured from local suppliers as a result of improvement in production productivity of parts, components and materials. *5*: Increase of in-house production ratio of parts, components, and materials, which previously imported from non-European suppliers. *6*: Entry of new Korean suppliers of relevant parts and materials into the EU. *7*: Switch of suppliers from non-EU member states to Korean capital-affiliated suppliers in the EU. *8*: Switch of suppliers from non-EU member states to firms other than Korean capital-affiliated parts and components suppliers in the EU. *9*: Considering the wishes or laws of governments of host countries.

per cent (eight companies) reported the existence of such ratios, while 38 per cent (five companies) stated they had no such ratios.

Concerning sources of material supply, a total of 16 companies gave answers. Of these, 69 per cent (11 companies) said that parent companies were major suppliers. Only three companies said that they got most material in the countries where they were operating. One company (Samsung Electronica Espanola S.A) obtained materials fairly evenly from the parent company (30 per cent), local suppliers (30 per cent) and third-party suppliers (40 per cent).

Relating to local content ratios of parts, components and materials, 16 companies again replied. Of these, 69 per cent (11 companies) said that the local content ratios of parts, components and materials which they used were increasing. Only two companies replied that the ratios were decreasing. Three companies said that the local content ratio had remained unchanged.

The major reason for the increase of local content ratio was the on-time delivery of local suppliers with the average value of '2.8', as Table 8.22 shows. This meant that 'on-time delivery of local suppliers' was an important reason for the increase in local content ratios of Korean consumer electronics firms in the EU. Besides the reasons given, three companies said that they used more locally produced parts and components because of local content or their localisation strategy. The two major reasons for the decrease of local content ratio were the price increase of locally produced parts and local raw materials, and the deterioration of quality of locally produced parts and locally supplied raw materials.

8.3 Conclusion

Korean FDI activities in the EU have increased steadily since 1986 when the SEM project was introduced by the EU. As of the end of 1994, there were 46 cases of Korean manufacturing firms' FDI in the EU. Most FDI of Korean manufacturing firms in the EU has been made by consumer electronics firms. In fact, 26 out of 46 cases of Korean FDI in the manufacturing sector were made by consumer electronics firms. This could happen because first, consumer electronics products, which have been one of the major Korean exporting items to the EU, have been most seriously restricted by EU's various trade and non-trade barriers. Second, Korean electronics firms have a more competitive edge, relatively speaking, in advanced foreign markets than other Korean industries, owing to the strong support of the Korean government. The Korean government designated the electronics industry as a 'strategic export industry' and enacted the Electronics Industry Promotion Act in 1969, entitling the industry to various government subsidies (Bark 1991). In addition, as Park (1985: 29) explained, 'the increasing manufacturing investment in advanced countries has been placed primarily in capital-and technology-intensive fabrication and assembly'.

The most interesting and important thing which the survey and interviews revealed is that FDI decisions of Korean consumer electronics firms in the EU are a 'defensive' and 'forced' response to pressures external to the firms. This is quite important, because much managerial literature or the theory of the firm argues about internal factors like transaction costs etc.[12] As analysed previously, the first motive for advancing into the EU was to carry out a so-called 'globalisation' strategy, although three of the five principal motives given were related to avoiding the protectionist measures already taken by the EU or those that might be taken in the future. The interviews with managing directors of Korean consumer electronics firms in the EU, however, revealed that the Korean firms had started feeling the necessity for a so-called globalisation strategy because of their growing concern about protectionism arising in the EU as a result of market integration. It may be, then, that Korean consumer electronics firms started local production within the EU because they were worried about protectionism arising in the EU from the market integration which was a major external pressure on Korean firms.

In particular, according to the results of the survey, FDI decisions of these companies were made by themselves, not by the Korean government, which is quite different from the usual pattern of relations between government and business in Korea. All 16 responding companies[13] answered that, first, they themselves felt the necessity of making FDI in the EU and decided to start local production within the EU.

Considering the above points, as Han (1992: 27) insisted, the conclusions to be drawn from the experience of foreign operations of Korean consumer electronics firms in the EU apply only to investments in developed countries by firms from LDCs, where their export markets are perceived to be under threat from protectionism and where the firms can bear financial losses of their FDI projects for five years or more.

The survey also revealed that for Korean companies to become more successful in local production in the EU, a number of changes would have to be introduced. Of these, one of the most important and urgent was the lack of local R&D centres, to improve success in local production in the EU, Korean consumer electronics firms should either set up their own local R&D centres or make joint ventures with local firms or university research centres in the countries in which they are operating, mainly in order to manufacture products locally according to local consumers' needs.

9

CONCLUSION

I have two main objectives in this final chapter. One is to recapitulate briefly the results of this research along with its principal findings and their implications. The other objective is to suggest several topics for future research that can build upon the work done for this book.

9.1 Recapitulation

The FDI of Korean firms, especially consumer electronics firms in the EU, has increased dramatically since 1987. Then, the Korean government did not take the initiating role in the FDI decision-making procedure of Korean consumer electronics firms in the EU. This is exceptional in Korea where the economy has developed very rapidly under the strong initiatives of the government.[1]

The main questions in the book are: 'Why did Korean consumer electronics firms particularly invest so heavily in the EU after 1987?' and 'Why did the Korean government eventually end up encouraging Korean consumer electronics firms to invest in local production in the EU rather than within the domestic Korean economy?'

To address the questions, a model of FDI decision-making of Korean consumer electronics firms in the EU was designed which considered various internal and influencing external factors which could affect the FDI decision-making procedure of Korean firms. Hypotheses derived from the model were evaluated using empirical data gained from a survey of 26 Korean consumer electronics firms operating in the EU as of the end of 1994 and personal interviews with senior managers of nine of these firms.

The answers to the questions are related to the national competitiveness of Korea. Several leading academic authorities on the subject, such as Porter (1990a: 2, 84; 1990b: 71–3) and Krugman (1994: 34), assert that the national competitiveness of a country is decided by the comparative productivity of firms within the country. However, this explanation is not enough to explain the Korean case, where national competitiveness is determined by the competitiveness of Korean firms in the international market. Given the small size of their domestic market, Korean firms are obliged to sell their products in foreign countries. Recent changes in the Korean market, such as the appreciation of the won and wage increases resulting from labour unrest, forced Korean consumer electronics firms to expand sales in foreign countries. This could be implemented by exports and/or by producing locally via FDI. In the mid-1980s, the EU Commission started to restrict exports of Korean consumer electronic firms by various tariff and non-tariff barriers. This new protectionism gravely threatened Korea's major exports to the EU. In addition, in 1986 the EU announced its '1992 single European market' plan, which, in 1986, was often interpreted by non-EU member states as a protectionist move. In response to these changes, Korean consumer electronics firms had little option but to invest in local production in the EU if they were to keep and if possible expand their market shares. In other words, Korean consumer electronics firms made FDI in the EU not for offensive reasons but for defensive reasons. Protectionism in both the USA and Japan had increased the importance of the EU as a place for doing business for Korean firms. The EU has emerged as one of the most attractive places for Korean firms to do business.

When Korean consumer electronics firms had to adjust to the changing external environment, the government had little choice but to follow the demands of the firms. This suggests a weakening in the traditional initiating role of the Korean government in its relations with firms. Instead, the firms themselves discovered the necessity of overseas investment and asked the Korean government for various kinds of support. The Korean government had no choice but to support consumer electronics firms when they encountered difficulties because it recognised that national competitiveness depended on them. In other words, the Korean government knew the competitiveness of firms were very important for national competitiveness. In addition, the Korean government, having adopted its so-called 'picking winners' strategy, would have found it difficult not to support the Korean firms. In the 1970s, the Korean government had deliberately chosen the so-called six 'strategic' industries – electronics, steel, non-ferrous metal, general machinery, shipbuilding and chemicals – to promote the Korean economy. The Korean government had provided these six 'strategic' industries with various privileges, such as preferential credits, both in terms of availability and cost preference, and other tax incentives, including investment tax credits, accelerated depreciation allowance and tax holidays (Leipziger and Petri 1989). The success of Korean firms in these selected 'strategic' industries, including the consumer electronics firms, has, according to both Porter (1990a: 84; 1990b: 71–3) and Krugman (1994: 34), enhanced the national competitiveness of Korea and contributed to the higher standard of living of its people. Given this importance, the interest of the Korean government is clear in supporting Korean consumer electronics firms when they encountered obstacles (especially anti-dumping decisions) to their exports to the EU.

In addition, the following major findings of the research, support the main hypotheses in the book:

1 The EU is very important for the Korean economy because of the internal and external market conditions facing Korean firms and the attractiveness of the EU market itself. The export-oriented development strategy of the Korean government, the internal nature of its trade, such as unbalanced trade relations (overconcentration on the relations with the USA and Japan) and labour unrest in 1987 and 1988, made the Korean economy heavily dependent on external markets, with the result that it is particularly vulnerable to changes in conditions of its external markets. Given this situation, the rising protectionism of the USA and Japan made the Korean government feel a strong necessity to diversify its trade relations. In this context, the EU emerged as an important future Korean trading partner.

   The sheer size of the EU market is by itself very attractive to non-EU countries. As of the end of 1995, the EU had a combined population of about 334 million, which exceeds that of the USA (263 million) and Japan (125 million). It had a GDP of US$ 7,082.7 billion in 1995, exceeding that of the USA (US$ 7,237.5 billion) and that of Japan (US$ 5,135.2 billion). The investment in the EU also offers non-EU countries the gateway to market access or opportunities for FDI in Central and Eastern Europe because of geographical and cultural proximity, plus the availability of market information (Lee 1990: 3). In addition, the possibility that all these countries will become members of the EU is reasonably high, considering that they have declared their intention of seeking membership in due course (Michalski and Wallace 1992: 113).
2 The uncertainty of the EU's common commercial policy (trade policy) was the main reason why Korean consumer electronics firms started their local production in the EU. The EU's common commercial policies have been characterised by instability, discrimination and lack of transparency (Tsoukalis 1993: 288–90; Hiemenz *et al.* 1994: 56).
3 Non-EU member states perceived the adoption of the SEM as a policy that could possibly strengthen protectionism of the EU member states. This impression remains, despite the efforts of EU member states to correct such a protectionist image of the SEM.
4 From the case studies of the internationalisation history of 16 Korean consumer electronics firms operating in the EU, it is clear that these firms were motivated by their need to seek market security through a 'globalisation' strategy. Faced by various forms of external threats over time, along with increasing internal capabilities, the Korean consumer electronics firms started to invest in local production in the EU for defensive reasons. Their lack of any monopolistic firm-specific advantages

left them no other feasible option. The defensive motivations of firms were reflected both in the shift from the export via foreign buyers to the local sales mode, and in the shift from the local sales to the local production mode. In the latter case, the Korean consumer electronics firms tried to protect their export markets in the EU which were threatened by various trade/technical barriers of the EU.

5 These firms claimed that one of the factors determining their choice of location for local production within the EU was the various investment incentives offered by host governments in the EU; and that such incentives also affected decisions on the size and timing of their investments. They also claimed that the incentives offered by their home government in Korea was a further important factor. The research, however, is inconclusive on this point. It cannot be proved that they would not have made such investments without tax and other incentives. In a series of interviews, none of the senior managers of nine Korean consumer electronics firms in the EU were able to give a clear answer to the question of whether or not they would have invested in local production in the EU without various investment incentives.

9.2 Future research

There are some interesting topics for future research emerging from the current research. A first major topic for future research is on the 'macro' internationalisation process of firms originating from LDCs. This process will deal with the successive decisions of firms originating in what used to be called LDCs to invest in production in new target markets such as the Eastern European countries, the USA and China. Many of the hypotheses developed in this research may equally hold true for other protected markets, such as North America or Australia.

Research is also needed on the internationalisation process of firms from LDCs in the Central and Eastern European countries in comparison with that in the EU. There would be some differences between the EU and the Central and Eastern European countries, even though the degree of differences between these two is smaller than that between the EU and other regions. The results of such a survey could be helpful for any manufacturing firms originating from LDCs which are considering entering the Central and Eastern European countries. After the collapse of communist regimes in Central and Eastern European countries, these countries are emerging as important and attractive places for outsiders. As an example, Korean manufacturing companies have already started investing in production in these countries. As of the end of 1995, 10 Korean manufacturing companies invested in these countries.[2] Korean companies in these countries expect to increase their production both in order to gain market access and to benefit from lower labour costs.

Research on the internationalisation process of firms originating from LDCs in China could also be interesting from a political economy perspective. China has recently emerged as the largest host developing country for FDI flows. According to the UN (1994: 14), the stock of inward FDI from 1982 to 1992 in China was US$ 33,768. An analysis of the motivations and strategies of firms which have invested in local production there may modify existing theoretical explanations of international production which have so far been based on the experience of firms from OECD countries.

Another major topic for future research could be the analysis of the changing relations between home governments and firms in LDCs other than Korea as a result of the internationalisation of firms. In particular, the impact of the home governments' trade and industrial policies on the internationalisation strategies of individual firms has often been neglected in the existing literature. Is the Korean case typical or peculiar? By comparing the trade and industrial policies of other home governments and the strategies of their firms in increasing world market share[3] with the Korean case examined in this book, it may be possible to develop a more comprehensive and generalised analysis of relations between home governments and firms in LDCs in the general theory to explain the internationalisation of production.[4] Systematic research to investigate this issue will draw interest from concerned parties, such as governments of developed countries and major firms in developed countries, which are facing strong challenges from the firms originating from LDCs.

NOTES

1 KOREAN ECONOMIC DEVELOPMENT AND THE IMPORTANCE OF THE EU MARKET TO KOREAN CONSUMER ELECTRONICS FIRMS

1 All EC member countries have used the term European Union (EU) officially since January 1994 according to the results of the Maastricht Treaty. In this book I use the term 'European Union' in referring both to the pre- and post-1994 situation. However, I of course retain the term 'European Community' (or EC) where it appears in quotations and the bibliography.

2 Refer to Chapter 2.

3 In 1980, the Korean economy experienced a negative economic growth rate of 5.7 per cent which was the first decline since the introduction of economic development plans in 1962. The decline rate resulted from the structural problems associated with past Korean industrial policies. Refer to Leipziger (1988); Sakong (1993); Park (1986) and Collins (1990).

The industrial policies that made problems were a) the excessive investment in industry which brought the inflationary pressure exerted on the economy, b) the unbalanced investment which brought the unbalanced growth among firms and industries, and c) the excessive protection of heavy and chemical industries which weakened export competitiveness of South Korean industry.

The political instability that was one of the major causes for the negative economic growth started from the assassination of President Park on 26 October 1979. This was followed by the dramatic increase of the trade deficit (US$ 1,781 million in 1978 and US$ 4,396 million), the large decrease of GNP growth rate from 7 per cent increase in 1979 to 4.8 per cent decrease in 1980. Refer to Table 1.1 in Chapter 1.

4 According to Korea Trade Promotion Corporation (1990) and Major Statistics of Korean Economy (1991), the total export and import volume in 1986, 1989, 1993 and 1995 took 64.6 per cent, 58.9 per cent, 50.1 per cent and 57.4 per cent out of GNP, respectively. The total export volume in 1986, 1989, 1993 and 1995 took 33.8 per cent, 29.1 per cent, 24.8 per cent and 27.6 per cent out of GNP, respectively. The total import volume in 1986, 1989, 1993 and 1995 took 30.8 per cent, 29.8 per cent, 25.3 per cent and 29.8 per cent, respectively.

5 They are higher tariffs against imports, quotas on imports, anti-dumping action, and so on. Some of these restrictions are old ones and the others are new ones.

6 According to Ahn (1990: 204), the USA is by far the largest export market for Korea. In 1987, 38.7 per cent of Korea's export earnings came from trade with the United States. Korea also depended on the Japanese market for 17.8 per cent of her total export earnings in 1987. The EU accounts for 10–15 per cent of the exports of Korea since 1980.

7 The Korean won was appreciated against the dollar under pressure from the US. The official foreign exchange rates against US$ in 1986, 1987 and 1988 were 861.4 won, 792.3 won and 684.1 won, respectively. The official foreign exchange rates of Japanese yen against US$ in 1986, 1987 and 1988 were 168.52 yen, 144.64 yen and 128.15 yen, respectively (National Statistical Office, Republic of Korea 1993: 327).

8 According to the Ministry of Finance (June 1991), the aggregate land area of the peninsula is 220,000 square kilometres, of which 98,500 square kilometres lie south of the truce line established at the conclusion of the Korean War in 1953. Its terrain is mostly mountainous (about 70 per cent of the total area), particularly in the north and east. The population of the Republic of Korea was about 45.5 millions, or about 458.8 persons per square kilometre as of the end of 1996 (National Statistical Office, Republic of Korea 1997).

9 According to Provost (1956: 219–24), as of 1944, a total of hydro-electric power developed or partially planned amounted to 5,863,720 KW. Of this 85.8 per cent was in North Korea.

10 'The US military administration had decreed that the military demarcation line between the northern zone of USSR occupation and that of the USA in the south should be the 38th parallel of latitude. This choice, which was made with the minimum of forethought, had very little logic other than to divide Korea in approximately equal geographical portions and to ensure that Seoul was in the USA zone. This arbitrarily drawn line, however, was to provide the basis for the long-term division of Korea into two independent states' (Lewis 1988: 8). In fact, the occupation was intended to be a temporary arrangement lasting only until elections supervised by the UN could be held.

11 The general election was held only in South Korea because the USSR refused to co-operate with the decision of the UN Temporary Commission on Korea (UNTCOK) to implement the USA's solution. The USA solution was to hold the internationally supervised elections throughout the Korean peninsula with a view of the formation of a single government chosen from the resulting national assembly. However, the USSR had advocated reunification through the adoption of a power-sharing arrangement between the north and the south that they had used as a prelude to complete communist dominance in Eastern Europe.

12 According to the Korea Development Institute (1975), damage during the Korean War was estimated as follows: US$ 512 million of dwellings, US$ 475 million of public utilities, US$ 232 million of medical facilities, US$ 345 million of industrial facilities and US$ 422 million of others. The estimated total damage reached US$ 3 billion, including the US$ 1.3 billion incurred by social overhead capital such as transportation and communication facilities.

13 The overvaluation of the exchange rate was especially pronounced towards the end of the Korean War, when the currency was devalued from 6 to 18 won to the US dollar. With continuing rapid inflation, however, the effects of the devaluation wore off by August 1955, when a new exchange rate of 50 won to the dollar

was adopted. This rate remained in effect until January 1960, notwithstanding continued domestic inflation. The real exchange rate, estimated by adjusting the official exchange rate for changes in wholesale prices at home and abroad, fell to half of its 1949–50 value (Frank, Kim and Westphal 1975: 30–3).

14 During the eight years following the armistice between 1954 and 1961 wholesale prices rose more than fourfold, at an average annual rate of 22.2 per cent. In particular, prices skyrocketed during the three years from 1954 to 1957. Inflation was temporarily suppressed in 1958, due to the large-scale influx of aid as well as the sudden drop in grain prices, but it became serious again in 1960.

15 According to the Economic Bureau, Ministry of Foreign Affairs (1963), about six million refugees fled to the south during the Korean War.

16 The Korean economy grew at an average annual rate of only 4.4 per cent from 1954 to 1961.

17 The Chang government was collapsed by a *coup d'état* staged on 16 May 1961 by a group of military officers headed by General Park Chung-Hee.

18 Since the early 1960s, economic and military aid from the US has decreased sharply. First, the US net economic assistance of Korean GNP at current prices was 10.75 per cent in the period 1956–7, while 7.28 per cent in the period 1963–4 (Schreiber 1970: 16). US military aid to Korea also decreased dramatically from US$ 4,765 million in the period 1946–65 to US$ 2,026 million in the period 1966–70. The US military aid to Korea continuously decreased; US$ 2,120 million in the period 1971–5, US$ 187 million in 1976 and US$ 158 million in 1977 (Farley, Kaplan and Lew is 1978: 16–17).

As part of an agreement to normalise relations between Korea and Japan, the Korean government could also get a grant of US$ 300 million and loans worth another US$ 200 million, which were crucial for the implementation of the first five-year economic development plan (1962–6). Korean firms depended heavily on foreign technology. In the 1950s, the US was the major supplier of technology. However, Korean firms imported 59.3 per cent of total imported technology from Japan between 1962 and 1979 (13 cases in the period 1962–6, 58 cases in the period 1967–71, 10 cases in 1972, 17 cases in 1973, 12 cases in 1974, 23 cases in 1975, 30 cases in 1976, 50 cases in 1977, 69 cases in 1978 and 20 cases until June 1979). The percentage of US technology out of total imported foreign technology was 22.9 per cent for the same period (10 cases in the period 1962–6, 197 cases in the period 1967–71, 34 cases in 1972, 46 cases in 1973, 62 cases in 1974, 58 cases in 1975, 76 cases in 1976, 79 cases in 1977, 154 cases in 1978 and 64 cases until June 1979 (*Dong-A Ilbo* (newspaper) 26 September 1979)

19 'By 1971, the construction of basic industries, such as steel, fertilisers, cement, oil refining and electricity, and the social infrastructure, such as highways, harbours, and irrigations facilities, were either completed or underway. Thus, by the end of this period, Korea had made the transition from agriculture-to manufacturing centered economy' (Lee and Yamazawa 1987: 56).

20 The major elements of the exchange rate reform took place in 1964–5 and included a 61 per cent devaluation of the won (May 1964) and a unification of the exchange rate (March 1965).

21 Established in 1966, the Council was renamed as the Trade Promotion Conference in the early 1970s; from 1983 to 1986, it met on a bi-monthly basis, but since then no meetings have been held. The Conference was usually attended

by the President, Cabinet Ministers, heads of financial institutions and major export firms, and not infrequently by Korea's ambassadors in major export markets (see Rhee, Ross-Larsen and Pursell 1984).

22 The government also established a National Investment Fund (NIF) in 1974 to help entrepreneurs' long-term investment in the industries. The NIF which composed for funds from government subscriptions, various public funds, and savings from banking institutions, was used to finance procurement of land, fixed and working capital, and long-term export finance requirements.

23 Refer to Collins (1990), Leipziger (1988) and Park (1986).

24 A *chaebol* is a grouping of enterprises owned or controlled by a person or family; either alone or in combination with other *chaebol*s, they hold dominant positions in a number of sectors, including automobiles, electronic products, shipbuilding, chemicals and petrochemicals. These positions were built largely during the early to late 1970s, when the real growth rate of the five largest *chaebol*s was roughly triple that of GDP. The share of the top 30 *chaebol*s in total shipments in mining and manufacturing peaked at some 40 per cent in 1985, but has settled at around 35 per cent since 1985. Refer to Economic Planning Board (November 1991) and Jones (1987).

25 Refer to Chapter 4.

26 In the early stage of industrialisation (1962–72), Korea relied more on technologies imported through capital goods, contact with foreign buyers and experiences in foreign-owned firms; less significant were licensing arrangements or foreign direct investment.

With the development of its economy and the shift from unskilled to skilled, labour-intensive industries, Korea began heavily importing proprietary technologies through licensing agreements. The dramatic increase in the number of licences approved during 1972–80 period is both the cause and effect of this structural change in the economy.

'During the 1962–80 period, Japan and the United States were the two major investors in Korean manufacturing industries. There were 68 direct investments from the United States, amounting to approximately US$ 146 million, and 413 direct investments from Japan, amounting to approximately US$ 332 million' (Lee and Yamazawa 1987: 90).

27 The proximity between the two countries reduced various costs, such as shipping costs, and also meant that less time was needed for travel for industrial and technical assistance personnel and management. Of the two major suppliers of licences, Japan was the more important, as the number of licences from Japan was more than double that from the United States. In terms of fees paid, however, the difference was not so great. Korea paid approximately US$ 168 million to Japan for the entire period and US$ 141 million to the United States. The average fee for a Japanese licence was US$ 166,000, whereas an average American one was US$ 360,000, more than twice the cost of a Japanese licence, even though the inflow of both Japanese and American licences was clearly concentrated in the high-technology industries.

28 The Republic of Korea was a colony of Japan from 1910 to 1945. The legacy of the Japanese colonial period still remains vibrant in many fields, which makes the countries similar in some respects. In fact, all educated Koreans over 60 could speak and read Japanese, because they had gone to elementary school in Korea

but had been taught in Japanese, and many of the elite had, in addition, been educated at Japanese universities. Some Korean business leaders had developed their careers as relatively junior workers at Japanese firms which later became major Korean corporations (Steinberg 1988: 46)

29 To do this, the government reduced the growth of money supply to the target zone of between 11.5 per cent and 15.0 per cent (1983–6). The average annual growth rate of money supply during the fourth five-year economic development plan (1977–81) was 24.7 per cent. The government also regulated the fiscal policy strictly with expenditures growing only at 7.5 per cent annually from 1980 to 1985 in real terms.

30 Before 1987, when the government gave full freedom to the labour movement, the average wage level of Korean workers were relatively lower than those of advanced countries and Asian NICs. According to ILO (1993), in 1986 average wage in manufacturing sector in Korea was only US$ 334. In 1986, average wages in the manufacturing sector in the USA and Japan were US$ 1557 and US$ 1812, respectively. In the same year the figure in Singapore was US$ 448.

31 Refer to Chapter 2.

32 In the period 1986–91, 45 per cent of outflows (measured in terms of amount) have been directed to North America (32 per cent in the US, 12 per cent in Canada) (US$ 1,550,513,000), and 39 per cent to South East Asian and Middle East countries (US$ 1,368,809,000). In the period 1978–85, 34 per cent of Korean FDI (US$ 165,041,000) went to North America, while 30 per cent (US$ 146,777,000) went to South East Asian countries and Middle East countries (Bank of Korea 1992).

33 Refer to Chapter 2.

34 As Lindert and Kindleberger (1982: 177–8) explain, there are four types of economic integration: a free trade area, a customs union, a common market and full economic union. The European Union is an example of a common market in which members allow full freedom of factor flows (migration of labour or capital) among themselves, in addition to having a free trade area.

35 The article in *European Trends* is aptly titled 'Lowering the Drawbridge on Fortress Europe'.

36 Professor Lee Tao-Yuan (1990) insists that 'Fortress Europe' was very much on the minds of those on the outside at his paper prepared for a conference on 'The EC After 1992 – Perspectives From the Outside', organized by the WWZ, University of Basel, held at Schloss Ebenrain, near Basel, Switzerland, 22–4 August.

37 For example, the United States authorities have proposed that there should be agreement on GATT rules prohibiting the trade restrictive and distorting effects of trade-related investment measures (TRIMs), which are the conditions imposed by host countries on operations established by foreign investors. 'Other TRIMs should be subject to GATT dispute-settlement procedures, under which countries found to be adversely affected would have the right to take countermeasures of "equivalent commercial effect"' (Hayes 1990: Executive Summary IV–V).

38 Even though the Commission has in the past refused to accept assembly as a 'substantial process or operation' that confers origin, the European Court of Justice judged that the assembly operations of Brother typewriters in Taiwan were a kind of substantial process.

39 These populations were as of the end of 1995 (National Statistical Office, Republic of Korea, 1997).
40 According to the State Statistical Bureau, People's Republic of China (1990) as of the end of 1989, the population of China was 1111.91 million. According to the World Bank (1993b), as of the end of 1991, the population of the Former Soviet Union was 289.8 million.
41 Refer to Table 8.1 in Chapter 8.
42 Korea promised to extend US$ 3 billion in bank and tied loans to the ex-Soviet Union: US$ 1.8 billion in 1991 and US$ 1.2 billion from 1992 to 1993. But it suspended the loans after the dissolution of the Soviet Union after paying US$ 1 billion in cash loans and US$ 470 million in export credits. According to *The Korea Herald* (17 and 24 May 1992), the Commonwealth of Independent States informed Korean creditors (10 Korean banks) that it could not pay the interest on a US$ 1 billion cash loan which Korea extended in 1991 on 15 May 1992. Through the contact between the two governments from the 22 to 24 May, Russia promised Korea that it would take over payments guarantees for all aid, standing at US$ 1.47 billion, which had already been offered to the ex-Soviet Union before it had broken up.

2 OVERVIEW OF KOREAN FDI

1 This assumes actual transfer of funds – which is not always the case.
2 Agmon and Kindleberger (1977: 145), and Wells (in Wells and Louis 1983: 4) state that investment among the developing countries flows primarily from the larger or wealthier developing countries to smaller or poorer ones. Agmon and Kindleberger also add that small investment flows from the developing countries to the developed countries and 'horizontally' in the developing world as the result of special production innovation or vertical integration. The FDI stock of developing countries in 34 host countries was estimated at some US$ 50 billion in 1985, with around US$ 36 billion invested in developed countries and the balance in the developing world (United Nations Centre on Transnational Corporations 1990; Fujita 1990). Estimates on the basis of data available from 58 host countries put FDI stock of developing countries at US$ 109 billion in 1990.
3 Net investments mean the outward flows of investments minus return flows through disinvestments from earlier investments.
4 According to the Bank of Korea (1992c), the Korean total amount of investment projects in mining and forestry industries showed a 60 per cent and a 70 per cent decrease through 1983–4, respectively.
5 Refer to Table 1.1 in Chapter 1.
6 Refer to Chapter 3.
7 Refer to Table 1.1 in Chapter 1.
8 According to the Korea Exchange Bank (1989: 69), current account surpluses were: US$ 4.6 billion in 1986; US$ 9.8 billion in 1987; US$ 14.1 billion in 1988; US$ 5.1 billion in 1989.
9 According to the IMF (1991), the conversion rates of won per US dollar from 1983 to 1990 were 775.7 won, 806.0 won, 870.0 won, 881.5 won, 822.6 won, 731.5 won, 671.5 won and 707.8 won, respectively.

10 According to the International Labour Organisation (1992), the wages in manufacturing sector increased by 12.2 per cent, 8.1 per cent, 9.9 per cent, 9.2 per cent, 11.6 per cent, 19.5 per cent, 25.1 per cent, 20.2 per cent, and 16.9 per cent in 1983, 1984, 1985, 1986, 1987, 1988, 1989, 1990, and 1991, respectively.
11 Refer to Chapter 5.
12 Refer to Chapter 8.
13 From Korean government.
14 The terms 'Europe' and the 'EU' are used fairly interchangeably in the context of Japanese investment depending upon how the Japanese aggregate their data. Although Europe includes the former Soviet Union, Eastern Europe and the non-EU countries in Western Europe, for practical purposes Europe and the EU are identical when discussing Japanese FDI. This is because 93 per cent of Japanese FDI in Europe is in the EU (with another 4 per cent in Switzerland) (Thomsen and Nicolaides 1992: 20).
15 The yen has appreciated since 1985. Exchange rates of yen per US$ in 1984, 1985, 1986, 1987, 1988, 1989, 1990, 1991 and 1992 were 237.52, 238.54, 168.52, 144.64, 128.15, 137.96, 144.79, 134.71 and 126.65, respectively (National Statistical Office, Republic of Korea 1993: 202).
16 The 12th *Survey of European Operations of Japanese Companies in the Manufacturing Sector*, conducted by Japan External Trade Organisation (JETRO) (1996).

3 THE POLICY AND PERSPECTIVE OF THE KOREAN GOVERNMENT TOWARDS OUTWARD FDI

1 Refer to Table 4.5 in Chapter 4.
2 Foreign Exchange Control Act, Law No. 933, as last amended by Law No. 1920 (30 March 1967)
3 Cabinet Decree No. 382, promulgated on 19 January 1962, as last amended by Presidential Decree No. 12777 of 10 August 1989.
4 The Foreign Exchange Control Regulations were first enacted by the Minister of Finance Notice No. 361 on 27 November 1964.
5 At that time Part 15 was Part 16, which was entitled 'Overseas Investment'.
6 Euh and Min (1986: 149–68) grouped the historical changes in FDI regulations by the Korean government into four periods: the incipient stage (1968–April 1975), the stage of streamlining the FDI system (April 1975–October 1978), the stage of strengthening the supervision of FDI (October 1978–October 1980) and the stage of vitalisation of FDI (October 1980–).
7 Since the late 1960s, in fact, the Korean government has attached primary importance to the role of outward foreign direct investment in (1) securing access to raw materials, (2) expanding exports, and (3) promoting international cooperation with developed and developing regions (Kumar 1981: 68).
8 'Bottleneck in exports' means barriers in exports.
9 Overseas Investment Guidelines, Article 15–6, (Korean government 1979).
10 'Emigration funds' are money which people can bring with them when they emigrate to a foreign country.
11 Technology services work was still considered foreign direct investment until July 1989 (as amended by MOF Notice No. 90–6, 2 March 1990).

12 Announced by the MOF on September 1987 and as amended by MOF Notice No. 87–20, 26 December 1987.

13 Foreign Investment Deliberation Review Internal Rules, Article 5(3), as amended on 23 December 1987.

14 MOF Bulletin No. 72 (July 1989); construction firms must provide 100 per cent financing from domestic Korean sources while real estate projects may borrow up to 50 per cent outside Korea. See also *BOK Guide* (Bank of Korea April, 1989: 5).

15 *Foreign Investment Guide* (Bank of Korea April 1989: 6); FECR then Article 15–5 entitled 'Encouraged Foreign Investment' was deleted.

16 FECR Article 15–6, as amended by MOF Notice No. 88–20, 25 October 1988.

17 The changes were announced in December 1993, in February 1994, in October 1995 and in June 1996.

18 The government has been increasing the annual ceiling of overseas investment credits provided by the Korea Export-Import bank from 40 billion won in 1988, to 100 billion won in 1989, to 120 billion won in 1990, to 100 billion won in 1991 (Ji 1991: 6).

19 This was also relaxed from the beginning of June 1996. In fact, according to the new regulation, it was liberalised to purchase the real estate for rental business, the construction of a golf course and the construction of buildings which will be sold.

20 Unproductive industry normally means service industry.

21 Korea Export Insurance Corporation (KEIC) was founded on 7 July 1992. The legal ground for establishment of the KEIC was provided by the Export Insurance Law which was revised on 27 March 1981. As of the end of June 1997, KEIC had five branch offices in Korea and nine representative offices in the world.

22 According to the Korean government, they often refer to all communist countries or former communist countries as 'Northern countries' or 'Socialist countries' in order to make the perception about communist countries clear.

4 THE RELATIONSHIP BETWEEN GOVERNMENT AND BUSINESS IN KOREA

1 The official diplomatic relationship between the Republic of Korea and the EU started on July 24, 1963. However, the Delegation of the Republic of Korea in the EU was opened in February 1986, and the Delegation of the European Union in the Republic of Korea was opened in November 1989.

2 Refer to Chapter 8.

3 The close and co-operative relations (partnership) between the government and business in Korea can indeed only be understood by considering history. First, in 1945 when Korea got independence from Japan, the Korean government took over all state-owned properties from the Japanese, which included factories, transportation and communication systems, and much of the land. After the Korean War (1950–3), the Rhee Syng-Man government sold these properties to private citizens, some of whom were entrepreneurs and some who were not, at bargain rates, with little obligation to pay the actual worth. This co-operative relationship between the government and business continued after Park became President (Ogle 1990: 42).

4 The Special Law for dealing with Illicit Wealth Accumulation was passed in the Second Chang Myon regime which was toppled by the military *coup d'état* led by General Park Chung-Hee on 16 May 1961. The guideline of the Special Law which the new Park Military government issued on 28 May 1961 was: 'The purpose of this guideline is in accordance with administrative procedures and criminal laws to deal with those who illicitly accumulated wealth. They include civil servants, politicians, and high-ranking government officials who took advantage of their positions and their power in their private fortune-making. They also include those businessmen and profiteering commercial agents who accumulated wealth illicitly by associating with above-mentioned government officials or politicians' (Jones and Sakong 1980: 290).

5 This condition (donating the shares to the government) was rarely fulfilled.

6 Due to the announcing date of this decision of the government, this is called as '8/3 Decree'.

7 Commercial bank of Korea, Hanil Bank, Korea First Bank, Seoul Trust Bank and Choheung Bank were not privatised until February 1973, July 1981, September 1982, and May 1983, respectively.

8 Two development institutions are the Korea Development Bank and the Export-Import Bank.

9 The same situation prevailed in Taiwan (China), where all banks were owned by government.

10 According to the calculations of Jones and Sakong derived from *Flow of Funds Accounts in Korea, 1963–1974* (Bank of Korea 1976), over the entire period of 1963 to 1974, only 14 per cent of cash flows of manufacturing corporations came from new equity. Another 20 per cent was generated internally, but two-thirds came from borrowing. Of the borrowing, 53 per cent came from domestic banks and financial institutions, 29 per cent from foreign sources, and 19 per cent from miscellaneous sources, including the curb market.

11 'All loans from abroad had to be initiated by the EPB or gain its approval. Foreign investments were all channelled through the government out to selected Korean enterprises. This system, plus the proclivity to direct funds to firms, that had succeeded in the past helps account for the steady, but rapid expansion of a comparatively few firms to Chaebol status' (Ogle 1990: 43).

12 In order to finance their first five-year plan (1962–6), the planners turned to Japan. As part of an agreement to normalise relations between Korea and its former colonial master, the Japanese government provided a grant of US$ 300 million and loans worth another US$ 200 million. Private Japanese firms were persuaded to add another US$ 300 million in investments. This US$ 800 million along with continued aid from the US (US$ 12 billion from 1945–65) assisted South Korea through the first five-year period. The third plan (1971–6), for example, was financed in part by US$ 4.5 billion loans from foreign commercial and governmental sources. Direct foreign investments added another US$ 488.1 million. During the time of the next plan (1977–81) foreign loans tripled to US$ 13.1 billion and direct investments increased slightly to US$ 531.8 million. The fifth plan (1982–6) saw a small decrease, but nevertheless a very high rate, of investment figures more than doubled up to US$1.1 billion (Ogle 1990: 36).

13 According to the National Statistical Office (1993: 31), in 1980 all items of grain production decreased by 22 per cent and in particular rice production declined by 36 per cent.

14 The Korean government announced an ambitious intermediate-term import-liberalisation plan in 1978. However, the plan was suspended because of unexpected exogenous disturbances such as the second oil shock (Sakong 1993: 208).

15 Refer to Young (1989).

16 This movement was started by President Park Chung-Hee in the early 1970s to aid rural development. But under the President Chun Doo-Hwan government, his younger brother led the movement.

17 'Ilhae' is President Chun's pseudonym. After North Korea's bomb attack on President Chun during his official visit to Rangoon, Burma, which killed 11 Korean top government officials including seven ministers, President Chun set up this organisation in memory of those victims. This organisation focuses on issues of Korean reunification. Thus, the organisation was closely associated with the President.

18 The second group of financing and banking business includes short-term investment firms, special government-run financial organisation for farmers and fishermen.

19 I had interviews with Mr J.R. Suh of Korea Trade Centre London (a semi-governmental organisation), Mr Y.W. Lee of GoldStar Electronics (UK) Ltd., Mr D.W. Kim of Kolon Imperial Graphics (UK) PLC, Mr D.B. Lee of SKM Europe (UK) Ltd., Mr J.H. Lee of Samsung Electronics Manufacturing (UK) Ltd., Mr J.H. Kim of Saehan Media Ireland Ltd., Mr H.I. Jo of GoldStar Europe (Germany) Ltd., Mr. C.H. Kim of Daewoo France S.A.R.I., Mr H.G. Kim of Samsung Portugal Electro-Mecanicos, S.A. and Mr. S.Y. Choi of GoldStar Iberna Italy S.P.A.

I conducted survey with all 26 Korean consumer electronics manufacturers in the EU, of which 16 replied to the questionnaire.

20 For example, as discussed in Chapter 3, the Korean government relaxed the Foreign Exchange Control Regulations on Korean policy towards FDI.

21 Refer to Chapter 8.

22 They were Saehan Media Ireland Ltd., Trigem Computers (UK) Ltd., Inkel (UK) Ltd., Daewoo Electronics UK Ltd., Saronix Kony Ireland Ltd., and GoldStar Europe Gmbh.

23 They were Haitai Euro S.A., Samsung Electronicsche Bauelemente Gmbh, Samsung Corning Deutschland Gmbh, GoldStar Electric (UK) Ltd., Kolon Imperial Graphics PLC., GoldStar Iberna Italy S.P.A., and Samsung Electronica (Spain).

24 They were Samsung Electronics Manufacturing (UK) Ltd., SKM Europe Ltd., and Samsung Portugal Produtos Electro Mecanicos, S.A.

25 Refer to Table 2.13 in Chapter 2.

26 The specific industries, designated as 'strategic' industries were steel, non-ferrous metal, general machinery, shipbuilding, electronics and chemicals (Sakong 1993: 56).

27 Refer to Chapter 8.

28 Refer to Smith (1989) and Schott (1989).

5 THE IMPACT OF EU POLICIES ON FDI BY KOREAN ELECTRONICS FIRMS IN THE EU MEMBER STATES

1 When negotiations on the widening of membership started in the late 1980s, the original heading of the negotiations was 'European Economic Space'. But the title was changed to 'European Economic Area' because there was a suggestion that a space denotes emptiness (Mayes 1993: 28).

2 Refer to Chapter 1.

3 Refer to Chapter 5.

4 According to Vernon's model, 'products go through a cycle of initiation, exponential growth, slowdown and decline – a sequence that corresponds to the process of introduction, spread, maturation and senescence.' (Vernon 1971: 75)

5 The export volume which is big enough for making the country importing goods think or take some actions against the imports from the country.

6 'Contrary to what is commonly believed, trade barriers are neither a necessary nor a sufficient condition for foreign investment though they may be a motivating factor' (Thomsen and Nicolaides 1990: 10).

7 According to *Handbook of the International Trade and Development Statistics* (United Nations 1989: 422), the population of the EU in 1987 as 40.2 per cent of the total population of the developed countries and 6.4 per cent of the world population.

8 1988 figure at current prices and 1985 exchange rates. It represents 29.1 per cent of total GDP of OECD countries.

9 Refer to Chapter 6.

10 Refer to Chapter 6.

11 Trade volume with the EU member states, trade and non-trade barriers that the EU member states imposed on products of the country, and so on.

12 'The effects of the more efficient internal market, which come initially through lower costs (of transport and of easier access to lower-cost goods in other member countries) should lower EU export prices. Lower prices enhance EU competitiveness in third markets, hurting producers in other countries, but by the same token they improve other countries' terms of trade' (Page 1992: 22).

13 Philips is the biggest and oldest European consumer electronics firm, with operations in 60 countries. It is the world's largest manufacturer of colour televisions, ranked 31 in the Fortune International 500, well ahead of Thomson (rank 101) and Nokia (rank 307) (*Fortune* 26 July 1993).

14 Thomson is the only state-owned consumer electronics producer so it has tended to try to influence Brussels through the agency of the French government. Thomson has only recently emerged as a serious challenge to Philips as a European-based international consumer electronics firm.

15 The core businesses of Nokia – The Finish-owned conglomerate – were such areas as paper and electric cables until 1984. However, Nokia started to emerge as a giant consumer electronics firm from February 1984. In February 1984 Nokia bought Luxor of Sweden and Salora of Finland to become the leading TV producer in Scandinavia, manufacturing each year the same volume of sets as Ferguson in the UK. Nokia became the third force in the European colour TV industry, after Philips and Thomson, with a combined production of 2.5 million sets per year' (Cawson *et al.* 1990: 325).

16 At least non-EU countries' perspective, strategies of European consumer electronics firms are contradictory.

17 Four firm clusters are Philips–Grundig, Thomson, Nokia, and the Japanese firms. These firms are attempting to develop a coherent international strategy which treats Europe at least potentially as a single market, and which treats in different countries as part of the strategic whole (Cawson *et al.* 1990: 321).

18 Refer to Table 6.3 in Chapter 6.

19 Refer to note 8, Chapter 2.

20 Refer to Table 6.3 in Chapter 6.

21 Kolon Imperial Graphics PLC in the UK, SKM Europe Ltd in the UK, GoldStar Europe GmbH in Germany, GoldStar Iberna Italy SPA in Italy, Inkel (UK) Ltd, GoldStar Electric (UK) Ltd, Daewoo Electronics (UK) Ltd, Trigem Computers (UK) Ltd, and Samsung Electronics (UK) Ltd.

22 SaRonix Kony Ireland Ltd. in Republic of Ireland, Samsung Portugal Produtos Electronecanicos S.A. in Portugal, Samsung Elektronische Bauelemente in Germany and Samsung Corning Deutschland GmbH in Germany.

23 Haitai Europe S.A. in France.

24 They were Saehan Media Ireland Ltd in Republic of Ireland and Samsung Electronica Espanola S.A. in Spain.

25 Refer to Chapter 6.

26 Refer to Chapter 8.

27 The conditions which Korean firms consider are labour cost, incentives offered by local governments, development of infrastructure etc.

28 Refer to Chapter 8.

6 EU TRADE POLICIES (COMMON COMMERCIAL POLICY)

1. In the 1970s the Social Democratic Party (SDP) was the dominant party and the Christian Democratic Union (CDU) has been in power since 1982.
2. Before the signing of the Treaty of Rome, the Treaty of Paris of 1951 had set up the European Iron and Steel Community. For administrative and decision-making purposes, the Economic Community and the Iron and Steel Community were unified in 1967.
3. The Single European Act is silent about trade and competition policy. Its unspoken message is that, from now on, the Treaty of Rome will be applied in full. But the Act added to the Treaty of Rome articles 130f–q, which defined Union responsibility for promotion of EEC technological development. Some people are arguing that this justifies a more protectionist interpretation of the rest of the Treaty.
4. See *Official Journal of the European Communities*, L209, (Commission 1988b).
5. See *Official Journal, L282*, Vol. 32, (Commission 1989a).
6. See *Official Journal, L252*, Vol. 27, (Commission 1984b).
7. See *Official Journal, L35*, (Commission 1982).
8. See *Official Journal, L93*, (Commission 1968).
9. As of April 1989, the EU took 17 measures under Article XIX. Four measures were taken in 1970s and 13 in 1980s. Agricultural products account for almost two-third of all cases. In addition, four measures were related to imports of electrical or electronic products (GATT 1991: 99).

10. The first revision came in January 1980 following the negotiation through GATT of a new anti-dumping code. Further revisions occurred in July 1984, June 1987 and July 1988.
11. Council regulation No. 2423/88, Article 2:3a.
12. Council regulation No. 2176/84 (Commission 1984a) was amended by Council Regulation No. 1761/87. Council Regulation No. 2423/88 (Commission 1988b), Article 13:10 sets out the existing law. See also the sixth *Annual Report of the Commission on the Community's Anti-dumping and Anti-subsidy Activities* (Commission 1989b), pp. 12–13 for an explanation of this regulation.
13. Certain agriculture-based products, such as wine and spirits, and coal, are subject to specific duties. Tobacco, some fruit and vegetables, certain carpets and glass categories, and watches come under alternative tariffs.
14. Further imports under specific duties include certain salt categories and cinematographic films.
15. Imports of many agricultural products are subject to levies under the common agricultural policy either in place of, or in addition to, the fixed tariff rates.
16. Each line represents a product sub-group.
17. The original members of EFTA were Norway, Sweden, Iceland, Austria, Switzerland, the United Kingdom, Ireland and Portugal. The United Kingdom, Ireland and Portugal joined the European Union in January 1973. Finland and Lichtenstein joined EFTA subsequently.
18. The first Lomé Convention was negotiated in 1973 as a successor to the Yaoundé Conventions covering the former French dependencies. In addition to tariff preferences, the Lomé Conventions provide for special quotas for exports of sugar and beef from Asia, Caribbean and Pacific countries to the Union, economic aid and compensation for shortfalls of export earnings from both agricultural products and minerals.
19. The Union refused to include the Asian countries of the Commonwealth under the Lomé Conventions.
20. According to Pohl and Sorsa (1992), the GSP involves some 40,000 bilateral and EU-wide tariff rebates. Renewed annually, it extends preferential trade treatment to some 146 developing countries and territories (not including Eastern Europe and Taiwan). The effective beneficiaries of the GSP scheme, as of the end of 1992, were 66 developing countries in Asia and Latin America. Benefits under the GSP scheme are temporary and nonbinding. Since the mid-1980s the GSP system has been subject to substantial differentiation among the beneficiaries.
21. The GATT regards the EU as a country for these purposes.
22. Imports of many agricultural products are subject to levies under the common agricultural policy either in place of or in addition to the fixed tariff rates.
23. As of August 1990, France (71 cases) and Italy (48 cases) accounted for the lion's share of the measures among more than 120 import restrictions on industrial products under Council regulation No. 288/82 (Commission 1982).
24. See *Official Journal*, L 284 (Commission 1991b).
25. See *Official Journal*, L 287 (Commission 1992b).
26. In October 1992, some of the restrictions maintained by France and enumerated in the Annex to Regulation No. 288/82 were termed 'quantitative restrictions without restrictions on quantities'.

27. The Treaty of Rome lays down that the Council shall exercise its powers in matters of trade policy by 'qualified' (that is, weighted) majority vote. Each country is given a number of votes according to its economic weight (Hayes 1993).
28. Council Regulation (EEC) 204/68, *Official Journal*, L 303, (Commission 1968b).
29. As a subsequent step, the list of liberalised products was further refined by transferring it from detailed 'Nimexe' categories. Nimexe stands for nomenclature of goods for the external trade statistics of the Union and statistics of trade between member states.
30. Council Regulation (EEC) 288/82, *Official Journal*, L 35, (Commission 1982).
31. According to Möbius (1991) in 1991 there would still be 819 national restrictions for industrial products after the termination of the Eastern European ones. Many of the restrictions are not enforced. Most are maintained by France (312) and Italy (386), but only the United Kingdom and Germany have no national quantitative restrictions. In agriculture most quotas are applied seasonally or to only a few suppliers (EU COM(90) 194).
32. In the Akzo–Dupont case covered by section 337 of the United States Tariff Act 1930 and in a case concerning sound recordings in Indonesia. Section 337 of the United States Tariff Act went to a GATT panel whose report was adopted by the GATT council in 1989. In the sound recording case, the commission decided to terminate the investigation after a solution was found by the firms involved. (NCC 1993: 76)
33. Trade in coal and steel is formally the responsibility of the ECSC.
34. Small-screen CTV means TV sets with screens up to 16 inches.
35. The Maastricht Treaty strengthens the enforcement of EU legislation by enabling the Commission to assess fines on member states if the Court has confirmed a violation of treaty obligations.
36. The Treaty of Rome, Article 113, section 4.
37. American Glenda Rosenthal concluded that explanation in terms of the activities of elite networks appeared to throw most light on decision-making in the Union in *The Man Behind the Decisions: Cases in European Policy-making* (Rosenthal 1975).
38. The Article 113 Committee plays an important part in the making of trade policy. It is so called because a clause in Article 113 of the Treaty of Rome specifies that the Commission shall conduct trade negotiations with countries outside the Union 'in consultation with a special Committee appointed by the Council to assist the Commission in this task and within the framework of such directives as the Council may issue to it'. While the role of the Article 113 Committee is nominally advisory, in practice this committee, consisting of officials with detailed knowledge of the issues and of the aims of their respective ministers, is an important link in the policy-making process (Hayes 1993: 39).
39. For an earlier and more detailed assessment, that is focused on EU protectionism against DCs (Hiemenz *et al.* 1990) and Tsoukalis (1993: 288–90).
40. Refer to GATT (1991) and Hiemenz *et al.* (1990).

7 INCENTIVE SYSTEMS OF EACH EU MEMBER STATE

1 See (Commission of the European Communities 1989d).

2 Fiscal concessions include accelerated depreciation, preferential tax rates, tax exemptions and tax credits, and measures relating to social security contributions and investment reserves.
3 According to Yuill, Bachtler and Wishlade (1994: 439), Italy does not offer interest-related subsidies any more due to Law 488 of December 1992 which abolished 'Special intervention' in the Italian Mezzogiorno.
4 The Municipality Soft Loan awarded loans to municipalities at low (7 per cent) rates of interest to allow them to construct industrial buildings which were then leased or bought by specified clients at below market rates (Yuill, Bachtler and Wishlade 1992: 131).
5 Mezzogiorno is the name of a southern area in Italy. In this area, there are four main types of incentive available.
6 The United Kingdom has two categories of area for the British regional incentive package, they are 'Development Area' and 'Intermediate Area'.
7 The Regional Policy Memorandum, covering the period 1991–4 was approved by Parliament in November 1990 and came into force at the start of 1991.
8 Refer to Chapter 8.

8 THE FDI OF KOREAN CONSUMER ELECTRONICS FIRMS IN THE EU: EMPIRICAL SURVEY

1 Three companies are Daesung and Telemart Sa Daesung is still in test operation in Europe. Telemart Sa has invested only US$ 5000 in Europe.
2 Refer to Table 2.12 in Chapter 2.
3 Daewoo – 7.49 per cent, Lucky-GoldStar –10.37 per cent and Samsung – 8.69 per cent of the total value of Korean FDI in manufacturing sector in the EU between 1985 and 1992.
4 Nine senior managers were seven managing directors, one vice-president and one marketing manager.
5 The domestic trading was carried out by the Samsung Sanghoe established in 1938. A part of the trading was sales of Korean products, e.g., fruits and dried fish to Manchuria.
6 He purchased 85 per cent of Hungop (today's Hanil), nearly 50 per cent of Cho-Heung, and 30 per cent of the Commercial Bank of Korea.
7 See Horiuchi (1994).
8 Each management system is based on unique cultural heritage, which has been maintained by each nation. It is to attain high performance through effectiveness, although all management systems have a common feature in terms of achieving organisation's goal. Rationality and individualism are emphasised in the Western management system, particularly in the American management system. Impersonality is another feature. As an example, leadership style in the Korean management system is very authoritarian and paternalistic in comparison with the American management system, where leadership style is mostly participative. Loyalty is very important in the Korean management style. The American style emphasises profit in the mottoes of organisations, while human harmony and solidarity are considered important mottoes of Korean organisations (Chang and Chang 1994: 129–58).

9 Taihan Electric Wire Company, which was the third-largest firms in the Korean industry in the 1970s was established in 1974. But Daewoo only acquired this company in 1983. Daewoo Electronics Co. now prefers to say that the company was founded in 1974 rather than in 1983.

10 According to the interview with S.Y. Choi, president of GoldStar Iberna Italy SPA, they produce refrigerator in Italy. In this case, there are no Japanese companies producing refrigerators in Italy.

11 Refer to Table 2.12 in Chapter 2.

12 Refer to Porter (1985).

13 Out of 16 responding companies, six companies answered that the chairman/chief executives suggested the idea of the initial overseas investment. Seven companies answered that it was manager-level personnel who suggested the idea. In particular, three companies answered that the decision came almost simultaneously from chairman/chief executives and manager-level personnel.

9 CONCLUSION

1 The annual average growth rate from 1962 to 1988 was 8.7 per cent.

2 Samsung Electronics Hungarian R.T. and Elektro-3 Ltd started their operations in Hungary in 1989 and 1991, respectively. Samsung Calex Co. Ltd started its operation in Slovak Republic in 1992. Electromagnetica GoldStar S.R.L. and RODAE Automobile S.A. were established in Romania in 1991 and 1994, respectively. In Poland, Daewoo Electronic Manufacturing Poland S.P.Z. O.O. started business in 1993 and Daewoo Motor Poland Corp. and Polmot Electronics Co. Ltd did in 1985. In Slovenia, SEICOM started its production in 1994 (Bank of Korea 1996).

3 Refer to Sally (1992: 121–81). In the former Czech Republic, K & S Musical Instruments Ltd also started its operation in 1994.

4 Refer to Stopford, Strange and Healy (1992).

BIBLIOGRAPHY

Books and articles

Agmon, Tamir and Kindleberger, C. P. (1977), *Multinationals from Small Countries*, Cambridge, MA: The MIT Press.

Ahn, Chung-Si (1990), 'South Korea: Political Transition and Economic Policy-making in South Korea', in *Economic Policy-Making in the Asia-Pacific Region*, edited by John W. Langford and K. Lome Brownsey, Halifax, Nova Scotia, Canada: The Institute for Research on Public Policy.

Ahn, J. B. and Jung, Y. H. (1995), *GumYoung SilMyungJae SilSiEiu PyungGa Wha DaeCheck* [The Evaluation of Real Name using Bank Account System after its Operations in Korea], Seoul: Korea Tax Institute.

Alisov, N. V. and Valev, E. B. (eds) (1985), *Economic Geography of the Socialist Countries of Europe*, Moscow: Progress Publishers.

Amsden, Alice H. (1989), *Asia's Next Giant: South Korea and Late Industrialization*, New York: Oxford University Press.

Amsden, Alice H. and Kim, H. (1985a), 'The Role of Transnational Corporations in the Production and Exports of the Korean Automobile Industry', working paper 9-785-063, Division of Research, Boston: Harvard Business School.

Aoki, M. (1992), 'Ex-post Monitoring of Team Production by the Main Bank', Mimeo, Stanford University.

Armstrong, Harvey (1993), 'Community Regional Policy', in *The European Community and the Challenge of the Future*, edited by Juliet Lodge, London: Pinter.

Bachtler, John and Clement, Keith (1990), *The Impact of the Single European Market on Foreign Direct Investment in the United Kingdom*, London: Department of Trade and Industry, HMSO.

Balassa, Bela (1985), 'The Role of Foreign Trade in the Economic Development of Korea', in *Foreign Trade and Investment: Economic Development in the Newly Industrializing Asian Countries*, edited by Walter Galenson, Madison, Wisconsin: The University of Wisconsin Press.

Balasubramanyam, V. N. (1991), 'Putting TRIMS to good use', *World Development*, vol. 19, pp. 1215–24.

Balasubramanyam, V. N. and Greenaway, David (1993), 'Regional Integration Agreements and Foreign Direct Investment', in *Regional Integration and the Global Trading System*, edited by Kym Anderson and Richard Blackhurst, London: Harvester.

Bark, Taeho (1991), *The Korean Consumer Electronics Industry*, Working Papers on Trade Policy of the World Bank, Washington: World Bank.

Begg, I. (1989), 'The Regional Dimension of the '1992' Proposals', *Regional Studies*, 23(4), pp. 368–76, Oxford: Pergamon Press.

Bügenmeier, B. and Mucchielli, J. L. (eds) (1991), *Multinationals and Europe 1992: Strategies for the Future*, London: Routledge.

Brown, Drusilla K. (1988/9), 'Trade and Welfare Effects of the European Schemes of Generalized System of Preferences', in *Economic Development and Cultural Change*, 37 (4), pp. 757–76.

Castells, Manuel (1992), 'Four Asian Tigers With a Dragon Head: A Comparative Analysis of the State, Economy, and Society in the Asian Pacific Rim', in *States and Development in the Asian Pacific Rim*, edited by Richard P. Appelbaum and Jeffrey Henderson, Newbury Park, London and New Delhi: Sage.

Cawson, Alan, Morgan, Kevin, Webber, Douglas, Holmes, Peter and Stevens, Anne, (1990), *Hostile Brothers: Competition and Closure in the European Electronics Industry*, Oxford: Clarendon Press.

Cecchini, P. (with Catinat, Michel and Jacquemin, Alexis) (1988), *The European Challenge 1992: The Benefits of a Single Market*, Aldershot: Wildwood House.

Chang, Chan Sup and Chang, Nahn Joo (1994), *The Korean Management System*, Westport, Connecticut and London: Quorum Books.

Chaponnière, Jean-Raphael (1992), 'The NIEs Go International', prepared for the conference 'Europe, U.S. and Japan in the Pacific Region: Current Situation and Perspectives', sponsored by the LVMH Möet Hennessy Louis Vuitton Group and held at the INSEAD Euro-Asia Centre, Fontainebleau, France on February 14 and 15.

Cho, Dong-sung (1990), *HanKook Chaebol Youngu* [A Study of Korean Chaebols], Seoul: Daily Economic Newspaper Ltd.

Cho, Y. J. and Hellmann, T. (1993), *The Government's Role in Japanese and Korean Credit Markets*, New York: The World Bank.

Choi, Sung-No (1996), *1996 Nyun 30 Dae KiEup GipDan* [Analysis of the 30 Korean Big Business Groups for 1996], Seoul: Korea Economic Research Institute.

Chung, Kae H. and Lie, Harry K. (1989), 'Labour–Management Relations in Korea', in *Korean Managerial Dynamics* edited by Kae H. Chung and Hak Chong, Lee New York, Westport, Connecticut and London: Praeger.

Clement, K. (1988), 'Effizienzvorteile und Voraussetzungen der Binnenmarktliberalisierung, [Efficiency Advantages and Preconditions for Liberalisation of the Internal Market], *Wirtschaftsdienst*, VIII, pp. 424–8.

Cole, D. and Park, Y. C. (1979), 'Financial Development in Korea 1945–1978', *KDI Working Paper*, no. 7904, Seoul: KDI.

——(1983), *Financial Development in Korea 1945–1978*, Cambridge, Massachusetts: Harvard University Press.

Cole, D. C. and Lyman, P. N. (1971), *Korean Development: The Interplay of Politics and Economics*, Cambridge, Massachusetts: Harvard University Press.

Collins, Susan M. (1990), 'Lessons from Korean Economic Growth and from Political Instability in Korea', *American Economic Review*, May 1990.

Cooper, C. A. and Massell, B. F. (1965a), 'A New Look at Customs Union Theory', *Economic Journal*, 75, pp. 742–7.

——(1965b), 'Towards a General Theory of Customs Unions for Developing Countries', *Journal of Political Economy, 73*, pp. 461–76.

Dicken, Peter (1992), *Global Shift*, London: Paul Chapman Publishing.

Dunning, Joan H. (1986a), 'The Investment Development Cycle Revisited', *Welwirtschaftliches Archiv*, 122(4), pp 667–76.

——(1993), *Multinational Enterprises and the Global Economy*, Workingham: Addison-Wesley.

Economou, G., Tsaveas, N. and Politis, P. (1993), 'Greece: Shaping Factors', in *The European Challenges Post-1992*, edited by A. Jacquemin and D. Wright, Aldershot: Edward Elgar.

Emerson, M. (1988), 'The Economics of 1992: An Assessment of the Potential Economic Effects of Completing the Internal Market of the European Community', in *European Economy*, 35 (March), Commission of the European Communities.

Encarnation, Dennis J. (1992), *Rivals beyond Trade*, Ithaca, New York, and London: Cornell University Press.

Euh, Yoon-Dae and Min, Sang H. (1986), 'Foreign Direct Investment from Developing Countries: The Case of Korean Firms', *The Developing Economies*, 24 (2), pp. 149–68, Tokyo: Institute of Developing Economies.

Farley, Phillip J., Kaplan, Stephen S. and Lewis, William H. (1978), *Arms Across the Sea*, Washington, D.C.: The Brookings Institution.

Fujita, Masataka (1990), 'Transnational Corporations from Developing Countries', *The CTC* 30 (autumn), pp. 411–32.

Frank, C. R., Jr., Kim, Kwang Suk and Westphal, L. E. (1975), *Foreign Trade Regimes and Economic Development: South Korea*, New York: National Bureau of Economic Research.

Gatsios, C. and Seabright, P. (1989), 'Regulation in the European Community', *Oxford Review of Economic Policy*, 5(2) (summer), Oxford: Oxford University Press.

Gilman, Martin, (1981), *The Financing of Foreign Direct Investment*, London: Frances Pinter.

Grimwade, N. (1990), *International Trade and the Consumer, Working Paper 1: Consumer Electronics and the EC's Anti-dumping Policy*, London: NCC (National Consumer Council).

Guisinger, S. E., (1986), 'Host-country Policies to Attract and Control Foreign Investment', in *Investing in Development: New Roles for Private Capital*, edited by T. H. Moran, pp. 157–72, New Brunswick: Transaction Books.

Guisinger, S. E. and Associates (1985), *Investment Incentives and Performance Requirement*, New York: Praeger.

Han, Do-Sook (1996), *Uri NaRaeu HaeWaeGikJupTuJawha KwalRunJaeDoeu GaeSunBangHang* [Korean FDI and Improvement of FDI-related Regulations], Seoul: Korea Institute of Public Finance.

Han, Sun-Taik, (1992), *European Integration: The Impact on Asian Newly Industrialising Economies*, Paris: OECD.

Hasan, Parvez (1976), *Korea: Problems and Issues in a Rapidly Growing Economy*, Baltimore: Johns Hopkins University Press.

Hayes, J. P. (1990), *Foreign Direct Investment: Will the Uruguay Round make a Difference?*, London: The Royal Institute of International Affairs.

——(1993), *Making Trade Policy in the European Community*, New York: St. Martin's Press.

Hiemenz, Ulrich, Langhammer, Rolf J. and Schrader, Jög-Volker (1990), 'EC Trade Policies toward Developing Countries', in *Trade Policies toward Developing Countries*, edited by Yamazawa Ippei and Hirata Akira, Tokyo: Institute of Developing Economics.

Hiemenz, Ulrich, Grundlach, Erich, Langhammer, Rolf J. and Nunnenkamp, Peter (1994), *Regional Integration in Europe and its Effects on Developing Countries*, Tübingen, Germany: Mohr.

Hillyard, Mick (1992), 'The Former Soviet Union: The Question of Aid and Trade', background paper no. 284, 26 February, Economic Affairs Section, House of Commons Library Research Division.

Hindley, Brain (1988), 'Dumping and the Far East Trade of the European Community' in *The World Economy*, 11 (December), Amsterdam: Elsevier Scientific Publishing.

——(1989), 'The European Community: Drifting (or Steering) to Bilateralism', Processed.

——(1990), *Foreign Direct Investment: The Effects of Rules of Origin*, London: The Royal Institute of International Affairs.

Holmes, Peter (1991), *International Trade and the Consumer, Trade and Competition Policy: The Consumer Interest*, London: NCC (National Consumer Council).

Hong, Wontack (1979), *Trade, Distortions and Employment Growth in Korea*, Seoul: Korea Development Institute.

——(1993), 'Export-oriented Growth and Equity in Korea', in *Trade and Protecionism*, edited by Takatoshi Ito and Anne O. Krueger, Chicago and London: The University of Chicago Press.

Horiuchi, Toshihiro (1994), *Japanese Public Policy for Cooperative Supply of Credit Guarantee to Small Firms – Its Evolution since the Post War and Banks' Commitment, EUI* Working Paper no. 94/3, Florence: European University Institute.

Ito, Takatoshi (1992), *The Japanese Economy*, Cambridge, Massachusetts, and London: The MIT Press.

Jacquemin, Alexis, (1993), 'The International Dimension of European Competition Policy', *Journal of Common Market Studies*, 31 (1) (March) Oxford: Blackwell.

Ji, DongHyun (1991), 'Investment Opportunities in ASEAN', presented at the seminar on Korea–ASEAN trade and investment co-operation organised by Korea Institute for Industrial Economics and Trade (KIET) and held in Seoul, Korea from 22 to 25, May.

Johnson, Chalmers (1987), 'Political Institutions and Economic Performance: The Government–Business Relationship in Japan, South Korea, and Taiwan', in *The Political Economy of the New Asian Industrialism*, edited by Frederic C. Deyo, Ithaca, New York, and London: Cornell University Press.

Johnson, H. G. (1965), 'An Economic Theory of Protectionism, Tariff Bargaining, and the Formation of Customs Unions', *Journal of Political Economy* 73, pp. 256–83, Chicago: University of Chicago Press.

Jones, Leroy P. (1987), 'Jaebul and the Concentration of Economic Power in Korean Development Issues, Evidence and Alternatives', in *Macroeconomic Policy and Industrial Development Issues*, edited by II, Sakong, Seoul: Korea Development Institute.

Jones, L. P., and Sakong, II (1980), *Government, Business, and Entrepreneurship in Economic Development: The Korean Case*, Cambridge, Massachusetts: Harvard University Press.

Jong, Henkwouter de (1993), 'European Competition Policy: Goals and Achievements', in *The Structure of European Industry*, edited H. W. de Jong, London: Kluwer Academic.

July, Serge, 'Fortress Europe, 1993', *Newsweek*, 12 July 1993.

Jung, K. H. (1989), 'Business–Government Relations in Korea', in *Korean Managerial Dynamics*, edited by Kae H. Chung and Hak Chong Lee, New York, Connecticut and London: Praeger.

Jung, P. M., and Chang, Y. S. (1992), *The Economic Analysis on Korean Chaebols*, Seoul: Korea Development Institute.

Kim, Kwang, S. (1967), 'An Appraisal of the High Interest Rate Strategy in Korea', paper presented at Williams College, Williamstown, Massachusetts.

——(1991), 'Korea', in Vol. 2 of *Liberalizing Foreign Trade*, edited by Demetris, Papageourgious Michael Michaely and Armeane M. Choksi, Cambridge, Massachusetts: Basil Blackwell for the World Bank.

Kim, Seok Ki (1987), *Business Concentration and Government Policy: A Study of the Phenomenon of Business Groups in Korea, 1945–1985*, unpublished D.B.A. dissertation, Harvard University.

Kitamura, Toshihara (1991) 'Investment in Financial Services', in *Japanese Direct Investment in Europe* edited by M. Yoshitomi, I., Akimune, T., Kitamura, G., Kume, K. Matsumoto and K. Totsuka, Aldershot: Avebury.

Klein, L. P. and Hong, Pingfan (1993), "Fortress Europe' and retaliatory economic warfare', in *Protectionism and World Welfare*, edited by Dominick Salvatore, Cambridge, New York and Melbourne: Cambridge University Press.

Kojima, K. (1978a), *Direct Foreign Investment: A Japanese Model of Multinational Business Operations*, London: Croom Helm.

——(1978b), *Japanese Direct Foreign Investment: A Model of Multinational Business Operations*, New York: Praeger.

Kong, Byung-Ho (1995) *HanKook KyungJaeeue KulRuk LiDong* [The Shift of Power in the Korean Economy], Seoul: ChangHae.

Koo, H. and Kim, E. M. (1992), 'The Development State and Capital Accumulation in South Korea', in *States and Development in the Asian Pacific Rim*, edited by R. P. Appelbaum and J. Henderson, London: Sage.

Krugman, Paul (1994), 'Competitiveness: A Dangerous Obsession', *Foreign Affairs* 73(2) (March/April), New York: Council of Foreign Relations.

Kumar, Krishna (1981), 'Introduction', in *Multinationals from Developing Countries*, edited by Krishna Kumar and Maxwell G., Mcleod, Massachusetts: Lexington Books.

Kwak, Tae-Hwan, Patterson, Wayne and Olsen, Edward A. (eds) (1983), *The Two Koreas in World Politics*, Seoul: Kyungnam University Press.

Langhammer, Rolf J. and Sapir, André (1987), *Economic Impact of Generalized Tariff Preferences*, Thames essay no. 49, London.

Lee, Chung H. and Yamazawa, Ippei (eds) (1987), *The Economic Development of Japan and Korea*, New York: Praeger.

Lee, Ki-baik (1984), *A New History of Korea*, Seoul: IlChoKak.

Lee, Tsao Yuan (1990), paper prepared for a conference on 'The EC after 1992 – Perspectives from the Outside', organised by the WWZ, University of Basel, held at Schloss Ebenrain, near Basel, Switzerland, 22 to 24 August.

Leipziger, Danny M. (1988), 'Industrial Restructuring in Korea', *World Development*, 16(1), Oxford: Pergamon Press.

Leipziger, Danny M. and Petri, Peter (1989), 'Korean Incentive Policies towards Industry and Agriculture.' in *Korea's Political Economy: International Perspective*, edited by Jeffery G. Williamson and Vadiraj R. Panchamukhi, Honolulu: University of Hawaii Press.

Lewis, D. S. (ed.) (1988), *Korea: Enduring Division?*, London: Longman.

Lim, D. (1983), 'Fiscal Incentives and Direct Foreign Investment in Less Developed Countries', *Journal of Development Studies*, 19, pp. 207–12, London: Frank Cass.

Lindert, Peter H. and Kindleberger, Charles P. (1982), *International Economics*, Illinois: Richard D. Irwin.

Luedde-Neurath, Richard (1988), 'State Intervention and Export-oriented Development in South Korea', in *Developmental States in East Asia*, edited by Gorden White, London: Macmillan.

Macdonald, Donald Stone (1990), *The Koreans: Contemporary Politics and Society*, London: Westview Press.

Maeda, Masahiro and Hong, Ng Sek (1990), *A Report on An Academic Visit to Seoul, Korea on Labour Study*, Hong Kong: Centre of Asian Studies, University of Hong Kong.

Maldague, R, Leroux, P. and Gilot, A. (1993), 'Belgium: Shaping Factors', in *The European Challenges Post-1992*, edited by A. Jacquemin and D. Wright, Aldershot: Edward Elgar.

Mansfield, Edward D. (1993), 'Effects of International Politics on Regionalism in International Trade', in *Regional Integration and the Global Trading System*, edited by Kym Anderson and Richard Blackhurst, New York and London: Harvester Wheatsheaf.

Markusen, J. (1983), 'Factor Movements and Community Trade as Complements', *Journal of International Economics*, 14, pp. 341–56, Amsterdam: Elsevier Science.

Marques, A. (1992), 'Community Competition Policy and Economic and Social Cohesion', *Regional Studies*, 26(4), pp. 404–7, London: Carfax.

Mason, Edward S., Kim, MahnJe, Perkins, Dwight H., Kim, KwangSuk and Cole, David C. (1980), *The Economic and Social Modernization of the Republic of Korea*, Cambridge, Massachusetts, and London: Harvard University Press.

Mayes, David G. (1993), 'The Implications of Closer Integration in Europe for Third Countries', in *The External Implications of European Integration*, edited by David G. Mayes, London: Harvester Wheatsheaf.

Messerlin, Patrick (1990), 'Anti-dumping Regulations or Pro-cartel Law? The EC Chemical Cases', *PRE Working Paper series* 397, World Bank, 1990.

Michalski, Anna and Wallace, Helen (1992), *The European Community: The Challenge of Enlargement*, London: Royal Institute of International Affairs.

Michell, Tony (1988), *From a Developing to a Newly Industrialised Country: The Republic of Korea, 1961–82*, Geneva: International Labour Office (ILO).

Möbius, Uta (1991), *Einfuhr von Industrieprodukten. Chancen der Entwicklungsländer und der Osteuropäischen Länder auf dem Binnenmarkt 1992*, Berlin: Deutsches Institut für Wirtschaftsforschung, mimeo.

Montagnon, Peter, (1990), 'The Trade Policy Connection', in *European Competition Policy*, edited by Peter Montagnon, London: Pinter.

Moskowitz, Karl (1989), 'Ownership and Management of Korean Firms', in *Korean Mangerial Dynamics*, edited by Kae H. Chung and Hak Chong Lee, London, New York: Praeger.

Mundell, R. A. (1957), 'International Trade and Factor Mobility', *American Economic Review*, 47, pp. 321–35.

Nam, Chong-Hyun (1993), 'Protectionist U.S. Trade Policy and Korean Exports', in *Trade and Protectionism*, edited by Takatoshi Ito and Anne O. Krueger, Chicago and London: The University of Chicago Press.

Neven, Damien and Georges Siotis (1993), 'Foreign Direct Investment in the European Community: Some Policy Issues', *Oxford Review of Economic Policy*, 9(2).

Nevin, Edward (1990), *The Economics of Europe*, London: Macmillan.

Norall, Christopher (1986), 'New Trends in Anti-dumping Practice in Brussels', *The World Economy*, 9 (March), p. 7.

Nurkse, Ragnar (1966), *Problems of Capital Formation in Underdeveloped Countries*, Oxford: Blackwell, (1st Ed. 1953).

Ogle, George E. (1990), *South Korea: Dissent within the Economic Miracle*, London: Zed Books Ltd.

Oman, Charles (1993), *Globalisation and Regionalisation*, Paris: OECD Development Centre.

O'Sullivan, P., (1985), 'Determinants and impact of foreign direct investment in host countries', *Management International Review*, 24, pp. 28–35.

Padoa-Schioppa, T., (1987), *Efficiency, Stability and Equity: A Strategy for the Evolution of the Economic System of the European Community*, Report of a study group appointed by the European Communities, Brussels (mimeo).

Page, Sheila, (1992), *Some Implications of Europe 1992 for Developing Countries*, Paris: OECD.

Park, Sung-Jo, (1985), 'The Investment Strategy of Korean Transnational Companies: Quantitative and Structural Aspects and a Typology of Korean Management Systems. Is Korea Following the Japanese Strategy?' in *The Internationalization of Japanese Business* edited by Malcolm Trevor, Boulder, Colorado: Westview Press.

Park, Won A. (1989), *Korea's Exchange Rate Policy: Determinants of the Won/Dollar Exchange Rate*, Working Paper no. 8914, Seoul: Korea Development Institute (KDI).

Park, Y. C. (1985), 'Korea's Experience with External Debt Management', in *International Debt and the Developing Countries*, edited by G. Smith and J. Cuddington, Baltimore Johns Hopkins University Press.

——(1986), 'Foreign Debt, Balance of Payments, and Growth Prospects: The Case of the Republic of Korea', *World Development*, 14(8).

Pelkmans, Jacques (1987), 'The European Community's Trade Policy towards Developing Countries', in *Europe and the International Division of Labour: New Patterns of Trade and Investment with Developing Countries*, edited by Christopher Stevens and J. Verloren Van Themaat, London: Overseas Development Institute, Institute of Development Studies and Institute of Social Studies.

——(1993), 'ASEAN and EC–1992', in *External Implications of European Integration*, edited by David G. Mayes, London: Harvester Wheatsheaf.

Petrochilos, George A. (1989), *The Determinants and Types of Foreign Direct Investment*, Aldershot: Avebury.

Pohl, Gerhard and Sorsa, Piritta (1992), *European Integration and Trade with the Developing World*, Washington D.C.: The World Bank.

Porter, Michael E. (1985), *Competitive Advantage: Creating and Sustaining Superior Performance*, London: Collier Macmillan.

——(1990a), 'The Competitive Advantage of Nations', *Harvard Business Review*, pp. 73–93, March–April, Boston, Massachusetts: Harvard University.

——(1990b), *The Competitive Advantage of Nations*, London: Macmillan.

Provost, Shannon McCune (1956), *Korea's Heritage: A Regional Geography*, Massachusetts: University of Massachusetts Press.

Reddaway, W. B., Potter, S. T. and Taylor, C. T. (1968), 'UK Direct Investment Overseas: Final Report', University of Cambridge Department of Applied Economics, Occasional Papers no. 12, Cambridge University Press.

Rhee, Young H., Ross-Larsen, Bruce and Pursell, Gary (1984), *Korea's Competitive Edge*, Baltimore: The Johns Hopkins University Press.

Rock, Michael T. (1973), 'Cross-country Analysis of the Determinants of U.S. Direct Foreign Investment in Manufacturing in Less Developed Countries', Ph.D. thesis submitted to University of Pittsburgh; unpublished.

Roegge, K. (1988), 'Kann der EG-Binnenmarkt bis 1992 Realisiert Werden?' [Can the Single European Market be Achieved by 1992?], *Wirtschaftsdienst*, 68 (III), pp. 153–6.

Roemer, John E. (1975), *U.S.–Japanese Competition in International Market: A Study of the Trade-Investment Cycle in Modern Capitalism*, University of California, Institute of International Studies, Research Series no. 22, Berkeley.

Root, F. R. and Ahmed, A. A. (1978), 'The Influence of Policy Instruments on Manufacturing Direct Foreign Investment in Developing Countries', *Journal of International Business Studies*, 9 (Winter), pp. 81–93, Academy of International Business and University of South Carolina.

Root, Franklin (1990), *International Trade and Investment* (4th and 6th editions), Cincinnati: South-Western Publishing Co.

Rosenbluth, Frances McCall (1992), 'Government–Business Relations in Japan: The Case of Financial Deregulation', in *Modernization in East Asia: Political, Economic, and Social Perspectives*, edited by Richard Harvey Brown and William T. Liu, Westport, Connecticut, and London: Praeger.

Rosenthal, Glenda Goldstone (1975), *The Man behind the Decision: Cases in European Policy-making*, Lexington: Lexington Books.

Rugman, A. M. and Verbeke, A. (1991), 'Competitive strategies for non-European firms' in *Multinationals and Europe 1992: Strategies for the Future*, edited by B. Bürgenmeier and J. L. Mucchielli, London: Routledge.

Sakong, II (1993), 'Changed Development Strategy of the 1980s', *Korea in the World Economy*, Washington D. C. Institute for International Economics.

Sally, M. R. (1992), 'The Political Economy of French and German Multinational Enterprises in International Competition', unpublished Ph.D. dissertation, the London School of Economics and Political Science.

Sanchez, J. S. (1993), 'Spain: Shaping Factors', in *The European Challenges Post-1992*, edited by A. Jacquemin and D. Wright, Aldershot: Edward Elgar.

Sano, J. (1977), 'Foreign Capital and Investment in South Korean Development', *Asian Economics*, 23 (December).

Sapir, Andre, Buigues, Pierre and Jaquemin, Alexis (1993), 'European Competition Policy in Manufacturing and Services: A Two-speed Approach?' *Oxford Review of Economic Policy*, 9(2), Oxford: Oxford University Press.

Schott, Jeffrey J. (1989), 'US Trade Policy: Implications for US–Korean Relations', in *Economic Relations between the United States and Korea: Conflict or Cooperation?*, edited by Thomas O. Bayard and Soo-Gil Young, Washington, DC: Institute for International Economics.

Schreiber, Jordan (1970), *U.S. Corporate Investment in Taiwan*, Cambridge: Cambridge University Press.

Scott, Norman (1993), 'Protectionism in Western Europe', in *Protectionism and World Welfare*, edited by Dominick Salvatore, New York: Cambridge University Press.

Sheard, P. (1991), *Delegated Monitoring among Delegated Monitors: Principal–agent Aspects of the Japanese Main Bank System*, mimeo., Stanford University.

Shibusawa, M., Ahmad, Z. H. and Bridges, B. (1992), 'The Newly Industrializing Economies' in *Pacific Asia in the 1990s*, by M. Shibusawa, Z. H. Ahmad and B. Bridges, London and New York: Routledge for the Royal Institute of International Affairs.

Shin, Wook Shik, and Oh, Eugene J. (1990), 'Recent Developments in Korea's Foreign Investment' *Journal of World Trade Law*, 24 (December).

Sideri, Sandro (1990), *European Integration and the Third World*, Institute of Social Studies, Working Papers 86, The Hague.

Siggel, E. (1986), 'Protection, Distortions and Investment Incentives in Zaire', *Journal of Development Economics* (22), pp. 295–319.

Smith, C. (1979), 'A Change of Direction', *Financial Times* (Survey on South Korea), Monday 2 April.

Smith, Michael B. (1989), 'Korea as a US Trading Partner', in *Economic Relations between the United States and Korea: Conflict or Cooperation?*, edited by T. O. Bayard and Soo-Gil Young, Washington, D.C.: Institute for International Economics.

Song, Byung-Nak (1990), *The Rise of the Korean Economy*, New York: Oxford University Press.

Steinberg, David I. (1988), *The Republic of Korea: Economic Transformation and Social Change*, Boulder, Colorado Westview Press.

Steers, Richard M., Shin, Yoo Keun and Ungson, Geratdo R. (1989), *The Chaebol: Korea's New Industrial Might*, New York: Harper & Row.

Stopford, J. M., Strange, Susan and Henley, John S. (1991), *Rival States, Rival Firms: Competition for World Market Shares*, Cambridge: Cambridge University Press.

Stopford, J. M. and Dunning, J. H. (1983), *Multinationals: Company Performance and Global Trends*, London: Macmillan.

Strange, Roger (1993), *Japanese Manufacturing Investment in Europe: Its Impact on the UK Economy*, London and New York: Routledge.

Suh, Sang-Chul (1978), *Growth and Structural Changes in the Korean Economy, 1910–1940*, Cambridge, Massachusetts, and London: Harvard University Press.

Taylor, Robert (ed.) (March 1993), *Transnational Corporations from Developing Countries: Impact on their Home Countries*, New York: UN

Thies, G. (1991), 'Recent Development in State Aid Policy, Directorate-General for Competition,' Commission of the EC, Ninth Annual Competition Policy Conference of the American Chamber of Commerce, Brussels, 7 March.

Thomsen, Stephen and Nicolaides Phedon (1990), *Foreign Direct Investment: 1992 and Global Markets*, RIIA Discussion Paper no. 28, Royal Institute of International Affairs.

——(1992), *The Evolution of Japanese Direct Investment in Europe*, London: Harvester Wheatsheaf.

Todd, Daniel (1990), *The World Electronics Industry*, London and New York: Routledge.

Tsoukalis, Loukas (1993), *The New European Economy*, Oxford: Oxford University Press.

Vernon, Raymond (1966), 'International Investment and International Trade in the Product Cycle', *The Quarterly Journal of Economics*, 80, pp. 190–207.

——(1971), *Sovereignty at Bay: The Multinational Spread of U.S. Enterprise*, New York: Basic Books.

Viner, Jacob (1950), *The Customs Union Issue*, London: Stevens & Sons.

——(1989), 'Innovation Strategies for Firms in the Industrializing and Newly Industrialized Asian Countries', in *Global Business: Asian-Pacific Dimensions*, edited by Erdener Kaynak and Kam-Hon Lee, London and New York: Routledge.

Vogel, Ezra F. (1991), *The Four Little Dragons: The Spread of Industrialization in East Asia*, Cambridge, Massachusetts: Harvard University Press.

Wade, Robert (1990), *Governing the Market*, Princeton, New Jersey: Princeton University Press.

Weiss, Frank D., Heitger, Bernhard, Jütermeier, Karl Heinz, Kirkpatrick, Grant and Klepper, Gernot (1988), *Trade Policy in West Germany*, Kieler Studien, 217, Tübingen, Germany: Mohr.

Wells, Louis T. (1983), *Third World Multinationals: The Rise of Foreign Investment from Developing Countries*, Cambridge, Massachusetts: MIT Press.

Westphal, L. E. and Kim, Kwang Suk (1982), 'Korea', in *Development Strategies in Semi-Industrial Economies*, edited by Bela Balassa *et al.*, Baltimore: The Johns Hopkins University Press.

White, Gordon and Wade, Robert (1988), 'Developmental States and Markets in East Asia: An Introduction', in *Developmental States in East Asia*, edited by Gordon White, London: Macmillan.

White, L. J. (1993), 'Competition Policy in the United States: An Overview', *Oxford Review of Economic Policy* 9(2), Oxford: Oxford University Press.

Wilkins, Mira (1970), *The Emergence of Multinational Enterprise: American Business Abroad from the Colonial Era to 1914*, Cambridge, Massachusetts: Harvard University Press.

——(1974), *The Maturing of Multinational Enterprise: American Business Abroad 1914–1970*, Cambridge, Massachusets: Harvard University Press.

Woronoff, Jon (1992), *Asia's 'Miracle' Economies*, Armonk, New York: M.E. Sharpe.

Yanaga, Chitoshi (1968), *Big Business in Japanese Politics*, New Haven: Yale University Press.

Yoo, Young (1984), 'Industrial Policy of South Korea: Past and Future' in *National Industrial Policies*, edited by Robert E. Driscoll and Jack N Behrman, Cambridge, Massachusetts: Oelegeschlager, Gunn and Hain.

Yoo, Sangjin and Lee, Sang M. (1987), 'Management Style and Practice of Korean Chaebols', *California Management Review (Summer)*, pp. 95–110.

Youn, Jung-Suk (1989), 'Korean Democracy and the Limits of Political Engineering', in *Progress in Democracy: The Pacific Basin Experience*, edited by the Ilhae Institute, Seoul: The Ilhae Institute, pp. 143–61.

Young, Soo-Gil (1989), 'Korean Trade Policy: Implications for Korea–US Cooperation', in *Economic Relations between the United States and Korea: Conflict or Cooperation?*, edited by Thomas O. Bayard and Soo-Gil Young, Washington, D.C.: Institute for International Economics.

Yuill, D, Allen, K. Bachtler, J. and Wishlade, F. (1992), *European Regional Incentives, 1992–93*, 12th edition, London: Bowker-Saur.

———(1994), *European Regional Incentives, 1994–95*, 14th edition, London: Bowker-Saur.

Government and other official publications

Bank of Korea (1976), *Flow of Funds Accounts in Korea, 1963–1974*, Seoul.

——(1988), *Annual Report*, Seoul.

——(April 1989), *Foreign Investment Guide*, Seoul.

——*Principal Economic Indicators* (June); *Monthly Bulletin*, various issues (1966–1992), Seoul.

——(1991), *HaeWae TuJa TongKae YunBo* [Overseas Direct Investment Statistics Yearbook], Seoul.

——(1992a), *Economic Statistics Yearbook*, Seoul.

——(1992b), *HaeWae TuJa TongKae YunBo* [Overseas Direct Investment Statistics Yearbook], Seoul.

——(1992c) *HaeWhe TuJa HyunGi BupIn HyunWhang* [Current Status of Korean Affiliates Abroad], Seoul.

——(1995), *HaeWae TuJa TongKae YunBo* [Overseas Direct Investment Statistics Yearbook], Seoul.

——(1996), *HaeWhe TuJa HyunGi BupIn HyunWhang* [Current Status of Korean Affiliates Abroad], Seoul.

——(1997), *Economic Statistics Yearbook*, Seoul.

Centre for Economic Policy Research (CEPR) (1990), 'Monitoring European Integration: The Impact of Eastern Europe'.

Commission of the European Communities (1957), *The Treaty of Rome*, Article 113, section 4.

——(1968a) *Official Journal of the European Communities*, L 93 of 17/4/68.

——(1968b), *Official Journal of the European Communities*, L 303 of 18/12/68.

——(1981), *Principal Regulations and Decisions of the Council of the European Communities on Regional Policy*, Luxembourg: Office for Official Publications of the European Communities.

——(1982), 'Council Regulation (EEC) No. 288/82 of 9 February', *Official Journal of the European Communities*, L 35 of 9/2/82.

——(1984a), 'Council Regulation (EEC) No. 2176/84 of 30 July', *Official Journal of the European Communities*, L 201 of 30/7/84.

——(1984b), 'Council Regulation (EEC) No. 2641/84 of 20 September', *Official Journal of the European Communities*, L 252 of 20/9/84.

——(1984c), 'Council Regulation (EEC) No. 1787/84 of 19 June 1984 on the European Regional Development Fund', *Official Journal of the European Communities*, L 169 of 28/6/1984.

——(1988a), 'Council Regulation (EEC) No. 2052/88 of June 1988', *Official Journal of the European Communities*, L 185/9 of 15/7/88.

——(1988b), 'Council Regulation (EEC) No. 2423/88, *Official Journal of the European Communities*, L 209/1 of 8/88.

——(1989a), 'Council Regulation (EEC) No. 2886/89, *Official Journal of the European Communities*, L282 of 2/10/89.

——(1989b), *The Sixth Annual Report of the Commission on the Community's Anti-dumping and Anti-subsidy Activities*, Brussels.

——(1989c), *Guide to the Reform of the Community's Structural Funds*, Brussels.

——(1989d), *Official Journal of the European Communities*, L395 of 30/12/89.

——(1990), *Official Journal of the European Communities* L370/11 of 20/12/90.

——(1990a), *Panorama of EC Industry 1990*, Brussels.

——(1990b), *Official Journal of the European Communities* L 370/7, L370/23 and L370/24 of 31/12/90.

——(1991a), *The Eighth Annual Report of the Commission on the Community's Anti-dumping and Anti-subsidy Activities*, Brussels.

——(1991b), *Official Journal of the European Communities*, L284, Vol. 34, 12 October.

——(1991c), *Panorama of EC Industries 1991–92*, Brussels.

——(1991), 'The European Electronics and Information Technology Industry: State of Play, Issues at Stake and Proposals for Action', Brussels, 3 April.

——(1992a), *Panorama of EC Industries 1991–1992*, Brussels.

——(1992b), *Official Journal of the European Communities*, L 287, Vol. 35, 2 October.

——(1993), *Panorama of EU Industry 1993*, Brussels.

——(1994), *Panorama of EU Industry 1994*, Brussels.

Council for Economic Planning and Development, Republic of China (1996), *Taiwan Statistical Data Book*, Taipei.

Directorate-General of Budget, Accounting and Statistics, Republic of China (1996), *Statistical Yearbook*, Taipei.

Economic Bureau, Ministry of Foreign Affairs, Republic of Korea (1963), *Trade and Investment Guide to Korea*, Seoul.

EPB (Economic Planning Board), Republic of Korea (1980a), *Major Statistics of Korean Economy*, Seoul.

——(1980b), *Handbook of Korean Economy*, Seoul.

——(1986), *Major Statistics of Korean Economy*, Seoul.

——(November 1991), *Economic Bulletin*, Seoul.

EIAK (Electronic Industries Association of Korea) *Statistics of Electronic and Electrical Industries*, Seoul, each issue from 1984 to 1993.

General Agreement on Tariffs and Trade (GATT) (1947), Article VI.

——(1991) *Trade Policy Review: European Communities*, Vol. 1, Geneva.

——(1992a), *Trade Policy Reviews, Korea*, Vol. 1, October 1992. Geneva.

——(1992b), *Trade Policy Reviews, Korea*, Vol. 2, Geneva.

——(1992c), *Trade Policy Review: European Communities*, Vol. 1, Geneva.

——(1993) *Trade Policy Review, European Communities* Vol. 1, Geneva.

——(April 1993), *Trade Policy Review Mechanism: The European Communities*, Vol. A Document C/R/M/S/36A, Geneva.

HMSO (1982), 'House of Lords Select Committee on the European Communities, Session 1981–82 Seventeenth Report', *Internal Market: The Movement of Manufactured Goods within the Internal Market of the European Community*, 13 July, House of Lords Sessional Papers 1981–82 Vol. 11, London: HMSO.

ILO (International Labour Organisation) (1992), *Yearbook of Labour Statistics*, Geneva.

——(1993), *Yearbook of Labour Statistics*, Geneva.

ILT (Investing, Licensing and Trading Conditions Abroad) (1980), *Korea* (July), Seoul: Business International Corporation.

——(1983), *Korea* (June), Seoul: Business International Corporation.

IMF (International Monetary Fund) (1991), *Balance of Payments Statistics Yearbook*, Vol. 42. Part 1, New York.

——(1992), *Direction of Trade Statistics Yearbook*, New York.

——(1995), *Direction of Trade Statistics Yearbook*, New York.

JETRO (Japan External Trade Organisation) (March 1991), *The 7th Survey of European Operations of Japanese Companies in the Manufacturing Sector*, London.

——(1992), *The Eighth Survey of European Operation of Japanese Companies in the Manufacturing Sector*, London.

——(1995), *Analysis of Finance: Japan*, Tokyo.

——(1996), *The Twelfth Survey of European Operations of Japanese Companies in the Manufacturing Sector*, London.

Korea Development Institute (1975), *Korea's Economy: Past and Present*, Seoul.

Korea Exchange Bank (1989), *The Korean Economy* (9th edition), Seoul.

KFTA (Korea Foreign Trade Association) (1976), *Overviews on Import Restrictions of Major Industrialised Countries*, Seoul.

——(1981), *Overviews on Import Restrictions of Major Industrialised Countries*, Seoul.

——(1985), *Overviews on Import Restrictions of Major Industrialised Countries*, Seoul.

——(1987), *Overviews on Import Restrictions of Major Industrialised Countries*, Seoul.

——(1989), *Overviews on Import Restrictions of Major Industrialised Countries*, Seoul.

——(April 1992), *Overviews on Import Restrictions of Major Industrialised Countries*, Seoul.

——(1997), *Overviews on Import Restrictions of Major Industrialised Countries*, Seoul.

Korea International Trade Association (1997), *Ju Yo Mu Yurk Dong Hyang Ji Pyo* [Major Trading Index], Seoul.

Korean government (1962), *Cabinet Decree*, No. 382, Seoul.

——(1979), *Overseas Investment Guidelines*, Seoul.

——(1989), *Presidential Degree*, Seoul.

Korea Trade Promotion Corporation (1990), *World Trade Statistics of Major Economies*, Seoul.

MOF (Ministry of Finance), Republic of Korea (1961), *Foreign Exchange Control Act*, Seoul.

——(1964), *The Foreign Exchange Control Regulations*, Seoul.

——(1986), *MOF Notice No. 86–23*, Seoul.

——(1987), *MOF Notice No. 87–20*, Seoul.

——(1987), *Foreign Investment Deliberation Review Internal Rules*, Seoul.

——(1988), *MOF Notice No. 88–8 and 88–20*, Seoul.

——(1989), *MOF Bulletin Nos 72*, Seoul.

——(1990), *MOF Notice No. 90–6*, Seoul.

——(June 1991), *Investment Guide to Korea*, Seoul.

Ministry of Public Information, Republic of Korea (January 1966), *Korean Economy: About to Take Off*, Seoul.
NCC (National Consumer Council) (1991), 'International Trade and the Consumer, working paper 5: trade and competition policy: the consumer interest', London: NCC.
——(1993), *International Trade: The Consumer Agenda*, London: NCC.
National Statistical Office, Republic of Korea (March 1992), *Korean Economic Indicators*, Seoul.
——(1993), *Major Statistics of Korean Economy*, Seoul.
——(1995), *Major Statistics of Korean Economy*, Seoul.
——(1996), *Major Statistics of Korean Economy*, Seoul.
——(1997), *Major Statistics of Korean Economy*, Seoul.
OECD (Organization for Economic Cooperation and Development) (1989), *Investment Incentives and Disincentives: Effects on International Direct Investment*, Paris.
State Statistical Bureau, People's Republic of China (1990), *China Statistical Yearbook*, Beijing.
Toyo Keizai Newspaper Ltd (1993), *Toyo Keizai* (spring), Tokyo: Toyo Keizai Newspaper, Ltd.
UN (1989), *Handbook of the International Trade and Development Statistics*, New York.
——(1992a), *World Investment Directory, Vol. 1: Asia and the Pacific*, New York.
——(1992b), *World Investment Directory, Vol. III: Developed Countries*, New York.
——(1994), *World Investment Report: Transnational Corporations, Employment and the Workplace*, New York.
United Nations Centre on Transnational Corporations (UNCTC) (1990), *Non-conventional Transnational Corporations*, New York: United Nations, E/C. 10/1990/18), p. 30.
World Bank (1993a), *The World Development Report*, Washington, D.C.: World Bank
——(1993b), *Statistical Handbook 1993: States of the Former USSR*, Washington, D.C.: World Bank
——(1997), *The World Development Report*, Washington, D.C.: World Bank.

Newspapers and periodicals

Atlantic News, (1992), 22 January.
Business Korea (1988), *Samsung After 50 Years: New Leadership Amid Changing Times*, March, pp. 25–40.
Dong-A Ilbo [*Dong-A* Daily Newspaper] (1979), 26 September, Seoul.
——(1986), 7 January, Seoul.
European Trends (1989), No. 1, London. The Economist Intelligence Unit.
Financial Times (1990), 8 January, London.
——(1990), 16 May, London
Forbes (1988), *Samsung: South Korea Marches to a Different Drummer*, 16 May, pp. 84–9.
Fortune (1988), *The Fortune International 500*, 1 August, D1–D36.
——(1993), *The Fortune International 500, 26 July.*
——(1994), *The Fortune International 500*, 25 July.
——(1996), August.
GoldStar Company Report (1988), pp. 6–7, Seoul: *GoldStar* Group.

HanKook Ilbo [*HankKook* Daily Newspaper] (1992), 15 March, Seoul.
Journal of Common Market Studies (1990), 28(3), March.
Ju Gan Sisa Journal [Weekly magazine *Sisa* Journal] (1993), 8 July, pp. 62–3, Seoul.
Korea Herald (1992), 17 May, Seoul.
——(1992), 24 May, Seoul.
Korean Economic Report (1989), November, Seoul.
Mall Kyung Jae ShinMun [Daily Economic Newspaper] (1989), 19 May, Seoul.
NaeWae KyungJae Shinmun [*NaeWae* Economic Daily Newspaper] (1992), 30 July, Seoul.
News from Daewoo (1987–8), (winter), p. 67, Seoul: Daewoo Group.
——*Daewoo* (1994), August 1, Seoul: Daewoo Group.
Newsweek, (1993), 12 July, p. 14.
Samsung: Its Role and Activities as a General Trading Company: 1986–1987 (1987), pp. 52–3, Seoul: Samsung Group.
UK Life, (1994), 22 June, p. 2, London.
Wol Gan Chosun [monthly magazine *Chosun*] (1994), January, pp. 206–18, Seoul
Wol Gan Sin Dong-A [Monthly magazine *Sin Dong-A*] (1986), August, p. 403, Seoul.

INDEX